Mining Haul Roads

Mining Haul Roads

Theory and Practice

Roger J. Thompson
Mineravia Consulting, Kalgoorlie, Australia

Rodrigo Peroni
Department of Mining Engineering, Federal University of Rio Grande do Sul, Porto Alegre, Brazil

Alex T. Visser
Department of Civil Engineering, University of Pretoria, Hatfield, South Africa

CRC Press
Taylor & Francis Group
Boca Raton London New York

CRC Press is an imprint of the
Taylor & Francis Group, an **informa** business

A BALKEMA BOOK

CRC Press
Taylor & Francis Group
6000 Broken Sound Parkway NW, Suite 300
Boca Raton, FL 33487-2742

First issued in paperback 2020

Typeset by Apex CoVantage, LLC

ISBN-13: 978-1-138-58962-9 (hbk)
ISBN-13: 978-0-367-62060-8 (pbk)

Library of Congress Cataloging-in-Publication Data
Names: Thompson, Roger, 1961 May 9- author. | Peroni, Rodrigo, 1971- author. |
   Visser, A. T., author.
Title: Mining haul roads : theory and practice / authors: Roger J. Thompson,
   Rodrigo Peroni & Alex T. Visser.
Description: Leiden, The Netherlands : CRC Press/Balkema, [2019] |
   Includes bibliographical references and index.
Identifiers: LCCN 2018043781 (print) | LCCN 2018049806 (ebook) |
   ISBN 9780429491474 (ebook) | ISBN 9781138589629 (hardcover : alk. paper)
Subjects: LCSH: Mine haulage.
Classification: LCC TN341 (ebook) | LCC TN341 .T46 2019 (print) |
   DDC 622/.69—dc23
LC record available at https://lccn.loc.gov/2018043781

**Visit the Taylor & Francis Web site at
http://www.taylorandfrancis.com**

**and the CRC Press Web site at
http://www.crcpress.com**

Dedication

This book is dedicated to Prof Alf Brown, erstwhile Head of the Department of Mining Engineering at the University of Pretoria, for his vision and support which lead to the development of the philosophy and practice presented in this book. In the early 1990s, Prof Brown approached Alex to guide Roger, a young senior lecturer who was looking for a topic for PhD studies, in the field of road transportation in open-cast mines. At that time, mines generally used a rough and ready rule of thumb approach, which consisted of selecting a road design based on prior knowledge and experience, and if that did not work, additional or alternative materials would be used until a satisfactory result materialised. This approach did not lend itself to an understanding of an integrated haul road design process, where safety, performance impacts, economics and cost per ton, hardly entered into the equation.

This project is a result of the integration of operational aspects, planning/design and civil engineering applied to mining engineering, combining the areas of expertise of each one of the three authors. The contents of this book stem from the original research described above, together with extensive mine-site application and developmental work supported by our various industry research partners and clients. As such, the authors gratefully acknowledge the support received from mine sites and staff whose efforts to understand and improve the performance of mine haul roads contributed to many practical aspects of these notes.

We also wish to acknowledge the patience of our families, as performing research in mines scattered around the world often meant that we were away from home. This situation reminded Alex of what a former civil engineering student encountered:

His father was also a civil engineer, and when the student prepared to get married his mother had a heart-to-heart talk with the bride-to-be. She explained to her that being married to a civil engineer (or mining in this case) meant that she will often be lonely, particularly in the first year of marriage, as the husband would forever be traveling to projects. "And after the first year?" the future daughter-in-law enquired excitedly. Mother then replied that after the first year the situation will not change, but the new wife will then have grown accustomed to being lonely.

This little story explains why our families deserve a great "Thank you".

Roger, Rodrigo and Alex

Disclaimer

Whilst the authors and publisher have made every effort to ensure the accuracy of the information presented herein, no warranty is given in respect of its accuracy, correctness or suitability. It is the responsibility of the reader to evaluate the appropriateness of particular information and methods or guidelines presented herein, in the context of actual conditions and situations and with due consideration to the necessity to temper this information with site-specific modifications and any overriding regulatory requirements. The authors and publisher cannot be held responsible for any errors or omissions in the book and accept no liability for any consequences arising out of the use of the information presented herein.

Contents

Foreword

This book is the most definitive treatise on mining haul roads ever written. A bold statement you say. During an extensive career in large surface mine equipment design and operations, like a lot of mining engineers, most of my learning about haul roads was through some hard lessons. Without a doubt this text would have been one of my valued resources. However, it was my work in the development and operation of autonomous trucks late in my career that would introduce me to the authors.

I had the responsibility to develop how the autonomous trucks should operate and then explain it to really smart people who would create the truck. This obviously led to considerable discussions about what haul roads should be, what they actually were and what they could be. It was this work that caused me to cross paths with Roger J. Thompson and recognize that he was a leading expert in this subject area. Through my work at the Society of Mining, Metallurgy and Exploration I was in a position to review and select technical papers from contributors from around the world. This gave me opportunity to review work by the authors and appreciate their in-depth understanding of this subject.

There has never been a text that addresses the many facets of mining haul roads on such a scope. I was particularly pleased to see the authors address not just design and construction concepts but also how haul roads affect the productivity and the total cost of ownership of the operations. This text has a section for every aspect of haul roads any engineer would need and as the title suggests, from both the theoretical and the practical aspects. I believe this comprehensive approach is critical when addressing this subject and that is what endears me to this text.

I would encourage anyone learning about or working with haul roads to make this text their go-to reference on the subject.

James (Jim) Humphrey, P.E.
Decatur Illinois

James D. Humphrey has over 35 years of worldwide experience in the mining industry. His recent work in the development of an autonomous mining system has led to multiple patents. Humphrey is a Mining Engineer graduate of the University of Missouri – Rolla. Additionally, he is a Professional Engineer, an author, and a Distinguished Member of the Society of Mining, Metallurgy and Exploration.

Abbreviations and notation

AASHTO	American Association of State Highway and Transportation Officials
ADT	Articulated dump trucks
AHS	Autonomous haulage systems
ASTM	American Society for Testing Materials
BCM	Bank Cubic Meters
BCR	Discounted benefit-cost ratio
BDT	Bottom dump trucks
CAPEX	Capital expenditure
CBR	California Bearing Ratio (%)
CCM	Compacted Cubic Meters
CDF	Cumulative Damage Factor
DCP	Dynamic Cone Penetrometer
DEM	Digital elevation model
DS	Defect Score
ESWL	Equivalent Single Wheel Load
EVM	Empty Vehicle Mass, also known as Tare mass
GHG	Greenhouse gas
GVM	Gross Vehicle Mass
HV	Heavy vehicle
IoT	Internet of Things
IRI	International Roughness Index
IRR	Internal rate of return
KPIs	Key Performance Indicators
LCM	Loose Cubic Meters
LV	Light vehicle
LVDT	Linear Variable Differential Transformer
MDD	Multi-Depth Deflectometer
MMS	Maintenance Management System
MSR	Mobile Sizing Rig
N-value	Weinert N-value describing physiographical regions for weathering of natural road materials
NPV	Net present value
OEM	Original equipment manufacturer
OMC	Optimum moisture content (%)
OPEX	Operating expenditure

OTR	Off the road
QAQC	Quality Assurance and Quality Control
RACS	Road analysis control system
RCA	Root cause analysis
RCM	Road condition monitoring
RDS	Roughness defect score
RDSI	Rate of increase of RDS
RDSMIN	Minimum roughness defect score
RDSMAX	Maximum roughness defect score
RDT	Rear dump trucks
ROPS	Rollover protection structure (or system)
RPI	Road Performance Index
RR	Rolling Resistance
RRI	Rate of increase in rolling resistance from RRMIN
RRMIN	Minimum rolling resistance at (RDS) = 5
RT-MMS	Real-time maintenance management system
SADT	Single Axle with Dual Tyres
SAST	Single Axle with Single Tyres
SD	Sustainable development
SWP	Standard Working Procedures
TKPH	Tyre tonne-kilometre per hour
UAV	Unmanned aerial vehicle
USCS	Unified Soil Classification System
UVM	Unladen vehicle mass
VHMS	Vehicle health monitoring system
VIMS	Vehicle information management system
VOC	Vehicle operating costs
ZAVL	Zero Air Voids Line
1F, 2F, 2R	Those symbols are related to the gears used when operating equipment, 1F – 1st gear forward, 2R – 2nd gear reverse.

Symbols

a	=	tyre contact radius, in m
A	=	Hourly operating area (m²/h)
$b_0, b_1, . b_n$	=	the benefits of a project, typically the reduced transportation costs, in each year
BE	=	Braking efficiency (generally close to 100%)
BF	=	Material bulking factor
B_n	=	Savings in operating costs
B_s	=	Bucket size (m³)
Bw	=	Universal dozer blade width (m)
$c_0, c_1, \ldots, c_n$	=	the costs of a project, typically construction and maintenance costs, in each year
C_{et}	=	Cycle time of the excavator for each load (mins)
CP	=	Estimated production from dozer production charts (Lm³/h)
CSx	–	Maximum change in cross-fall (m/m)
d_{50}	=	Sieve size at which 50% of a material passes
d_{90}	=	Sieve size at which 90% of a material passes
d_b	=	Stopping distance (m)
d_d	=	Dozing distance (m)
D	=	Days since last maintenance
D_{Load}	=	Engine load (percent of gross power available)
DM	=	Days between last maintenance and minimum cycle defect score
DR	=	Dust ratio ($P_{0.075}/P_{0.425}$)
DRR	=	Drive reduction ratio
DSMIN	=	Minimum defect score in maintenance cycle
DSMAX	=	Maximum defect score in maintenance cycle
e	=	Super-elevation applied (m/m width of road)
E	=	Stiffness or resilient modulus of a material, typically in MPa
E_{ff}	=	Working efficiency
EP	=	Engine power (flywheel) (kW)
F_{susp}	=	force acting between the sprung and unsprung masses
F_{tyre}	=	force exerted on the unsprung mass m_u by the tyre
g	=	Acceleration due to gravity (m/s²)
GC	=	Grading coefficient
GR	=	Longitudinal grade (%), -ve upgrade, +ve downgrade
H	=	Vehicle age khrs (total engine hours)

h	=	Compacted layer thickness, usually in mm or m
h_l	=	Headlights height above ground surface (m)
h_1	=	Height of operator eye in cab above ground (m)
h_2	=	Height of object above ground, usually 0.15 (m)
HC_{haul}	=	Hauling cost per hour
HC_{mtce}	=	Hourly maintenance cost
HD	=	Haul distance (km)
i	=	Discount rate (%)
i_d	=	Headlight divergence (°)
Job_{Ef}	=	Motor-grader working efficiency
KT	=	Average daily tonnage hauled (as payload) (kt)
k_{tyre}	=	Tyre stiffness when considered as a simple linear spring
L	=	Length of vertical curve (m)
LB	=	Layback on horizontal curves (m)
LC	=	Labour costs (/1000km)
LDDD	=	Rate of defect score decrease immediately following last maintenance cycle
LDDI	=	Rate of defect score increase
L_e	=	Effective blade length (m)
L_o	=	Width of overlap (m)
LL	=	Liquid Limit (%)
$LP_{0.425}$	=	Percentage of *loose* wearing course material passing the 0,425mm sieve
LS	=	Bar linear shrinkage (%)
L_v	=	Layer volume (m³)
M	=	Wearing course material type
MD	=	Maximum deceleration (m/s²)
nc	=	Number of unrestricted hauling cycle after maintenance
N_{DH}	=	Number of dozer hours
N_{eh}	=	Number of excavator hours
N_{GH}	=	Number of operating hours of the motor grader
N_{TL}	=	Number of truck loads
N_{TH}	=	Number of truck hours
N_L	=	Number of loader bucket loads
NP	=	Non-plastic material in the liquid limit test
n_{passes}	=	Number of roller passes required to achieve compaction specification
OMC	=	Optimum moisture content (%)
P	=	Tyre pressure (kPa)
PC	=	Parts cost (/1000km)
P_D	=	Dozer production, measured in loose cubic meters (Lm³/h)
PKDST	=	Peak dust reading (×100 mg/m³) of the minus 10 micron dust fraction, measured by Hund Tyndalometer
P_x	=	Percent passing sieve size × mm, 0.075mm, to 37.5mm
P_{Gross}	=	Gross engine power (SAE J1995) (kW)
P_{compac}	=	Compaction production measured in loose cubic metres
PF	=	Plastic factor (Plastic Limit × $P_{0.075}$)
PFC	=	Dozer production factor corrections
PI	=	Plasticity Index (%)
PL	=	Plastic Limit (%)

PR	=	Power delivered as rimpull (kW)
ProdF	=	Productivity factor
Q_f	=	Haul truck fuel consumption (L/h)
R	=	Radius of horizontal curve (m), taken as the left lane for determining stopping distance
Rain	=	Mean annual rainfall (mm)
Rate	=	Watering rate in l/m^2
R_L	=	Road Length (m)
RP	=	Rimpull (kN)
RPM_{Max}	=	Maximum engine RPM
Rw	=	Compacted road width (m)
S	=	Operating speed (km/h)
S_d	=	Dual wheel centre to centre spacing (m)
SD	=	Stopping sight distance (m)
Sp	=	Shrinkage product
SP	=	Slightly Plastic material in the liquid limit test
SRR	=	Speed reduction ratio
Sup_{Lat}	=	Lateral superposition between motor-grader passes (m)
T	=	Torque (kNm)
TC	=	Total cost of hauling operation
T_L or T_E	=	For truck, time loaded or time empty (min)
T_T	=	Total truck cycle time (min)
T_{SL}	=	Spotting and loading time (min)
T_{SU}	=	Spotting and unloading time (min)
t_{mtce}	=	Unrestricted maintenance time
t_{uop}	=	Unrestricted operation time
$t_{p\text{-}r}$	=	Driver reaction *and* brake activation time (s)
TW	=	Tyre set wear rate (tyres consumed per 1000 km for a six-wheeled truck)
t_w	=	Truck wheel load (metric tonnes) in the CBR thickness design method
TYPE	=	Indicator for truck type
U_{min}	=	Coefficient of longitudinal deceleration (friction supply) or lateral friction
V	=	Vehicle speed (km/h)
V_F or V_E	=	Speed of laden (F) or empty (E) truck (km/h)
V_G	=	Average grader speed during the execution of the task (km/h)
VEH_R	=	Replacement cost of vehicle ($\times 10^{-6}$)
V_L	=	Compacted volume of the layer (m^3)
V_{TRUCK}	=	Volume capacity of the truck (m^3)
VOL	=	Hourly traffic repetitions on haul road
v_o	=	Initial vehicle speed (m/s or km/h)
v_f	=	Final vehicle speed (m/s)
VW	=	Volume of water to be added (or evaporated) per meter length of road (m^3)
V_{fav}	=	Favourable truck speed, with gradient (km/h)
$V_{\varepsilon tmax}$	=	Truck speed at maximum efficiency (ε_{tmax})
v_{uf}	=	unfavourable truck speed against gradient (km/h)
w	=	Truck width (across body, excluding mirrors) (m)
WHL	=	Number of wheels on truck
W_r	=	Width of roller or rolled path (m)

WSHEAR	=	Wind shear (mm/s.mm) under the truck
W_t	=	Weight on a single tyre (kN)
WB	=	Wheel base of truck (m)
Z_{CBR}	=	Layer thickness for a material with a given CBR (m)
Z_{ESWL}	=	Cover thickness for a given Equivalent Single Wheel Load (m)
Z_r	=	Road elevation (a measure of the defect dimensions) (m)
z_u	=	Unsprung mass displacement (m)
α, β	=	Constants representing labour unit costs and the proportion of parts costs
ε_{lim}	=	Limiting vertical compressive strain
ϵ_t or ϵ_r	=	Transmission or retarder efficiency
ΔG	=	Algebraic difference between two grades on either side of a vertical curve (%)
Θ	=	Grade of road (degrees) positive (+ve) downgrade
ρ_{dry}	=	Material dry density (kg/m^3)
μ	=	Poisson's ratio
$\mu\varepsilon$	=	Microstrain or strain times 10^{-6}
ω_f	=	Moisture content in field (borrow pit) %

Chapter 1

Introduction to mine haul roads

1.1 Importance of mine haul roads

Mining operations may be classified into surface mining, where the operations are unconstrained and open to the sky, and underground operations, where mining takes place in tunnels and galleries. In both situations roads play a key role as a typical mining operation consists of moving overburden (waste material) and ore from where the material was formed or deposited, to waste dumps or beneficiation plants. In surface mining, material movements are typically made by vehicles ranging from conventional on-highway trucks, road-trains (tractor-trailer combinations) and special purpose off the road (OTR) or off-highway dump trucks as shown in Figure 1.1; the current largest being over 600 t gross vehicle mass (GVM). In contrast, underground operations material transportation takes place by load-haul-dump units, delivering to an ore or waste-pass system which then either feeds underground haul trucks (typically 40–80 t payloads), or a conveyor or shaft system which delivers the material to surface. In both cases, but more especially surface mining applications, the haul road plays a critical and central role in safe, efficient and cost-effective mining.

What comprises a mine haul road has various definitions, depending both on the class of vehicle using the road and the wheel and axle load limits of the road design itself. Three basic types of haul road can be defined, based on application and the specific type of trucks associated with it:

- Road legal:
    - Sealed (paved) roads:
        - Conventional truck and truck-semi-trailer combinations
        - Single, double or triple trailer road trains, < 120 t capacity
    - Unpaved roads, additionally:
        - B-double, B-triple and Quad road trains (with powered trailer options), < 170 t capacity
- Off road (unpaved) and in-pit:
    - Small scale operations and civil construction – articulated dump trucks (ADTs), < 75 t capacity
    - Custom road trains with > 200 t haulage capacity
    - Rear-dump trucks (RDTs) and bottom-dump trucks (BDTs)

(a) On-highway truck

(b) Road-train *(Image courtesy of Rivet Mining Services)*

(c) Articulated dump truck (ADT) *(Image courtesy of Bell Equipment)*

(d) Bottom (or belly) dump truck (BDT) *(Image courtesy of Kress Corporation)*

Figure 1.1 Surface mining haulage options.

(e) Rear dump truck (RDT)

Figure 1.1 (Continued)

In the case of surface mining, economy of scale and the increase in haul truck payload, as shown in Figure 1.2, have so far seen the RDT ultra-class truck (> 220 t payload, > 380 t GVM) population rise significantly (Gilewicz, 2006). With this increasing size, haul road performance can be compromised, resulting in excessive total road-user costs; often seen directly as an increase in cost per ton hauled, but also indirectly as reduced production rates and vehicle and component service life. Especially with the larger haul trucks now predominant in surface mining, it is advantageous to consider the haul road as an asset, similar to the vehicles that use the road.

Consequently, the road design and its performance and maintenance management processes should be integrated with the process of mine haulage systems management. Figure 1.3 illustrates the evolution of haul road design and management in terms of an evolving asset management system.

In the early 2000s, the concept of autonomous trucking (driverless RDTs) has moved from prototypes to production-ready applications. In this scenario, the operating performance of the haul road will become 'mission critical' to the overall success of autonomy in mining. Rapid deterioration of haul roads will require costly remediation, human intervention and significant, albeit temporary, changes to operating procedures, to accommodate these types of events. With autonomous trucking, vehicle path wander is minimal and the road will be subject to high channelised wheel loads over a limited area, without the wheel-path variations often encountered with conventional trucking.

Many concepts from highway engineering can be adapted to the design, construction and management of mine roads. However, significant differences in applied loads, traffic volumes, construction material quality and availability, together with design life and road-user cost considerations, indicate that a custom-made design solution is required. Underperformance of a haul road will impact immediately on operational safety and cost-efficiency, since up to 50% of the total costs of surface mining can be ascribed to haulage costs and mining is dependent on well-designed, constructed and maintained haul roads (Thompson and

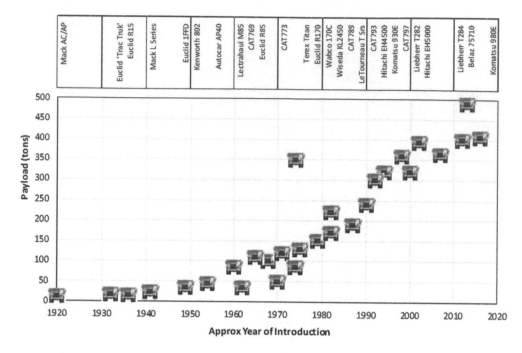

Figure 1.2 Trend in truck payload capacity for selected rear dump truck models from 1920s onwards.

Visser, 1999). Most mine operators will agree that a strong relationship exists between well-constructed and maintained roads and safe, efficient mining operations. However, mining roads are often seen as 'non-core' components of the mining process and, as a result, mining operations may not give due or targeted consideration to appropriate road design standards, construction, operational and maintenance management issues (Kecojevic *et al.*, 2007).

Where design standards and operational management input is lacking and an empirical design approach used, safe, economically optimal roads eventually result – but the learning curve is steep, slow and inherently costly and hazardous. This approach does not lend itself to an understanding of the road design process and more importantly, if haul road performance or safety is sub-standard, identifying the underlying cause of the unsafe condition and/or design deficiencies which contribute to poor performance or an accident (as a root-cause or associated factor), is problematic. An ad-hoc or empirical approach to haul road design is generally unsatisfactory because it has the potential for over-expenditure, both on construction and operating costs, arising due to the following:

i. over-design and specification, especially in the case of short-term, low-volume roads where the effect of rolling resistance, although minimised, does not contribute significantly to reducing total road-user costs across the mine's network of roads due to the higher initial construction cost; or

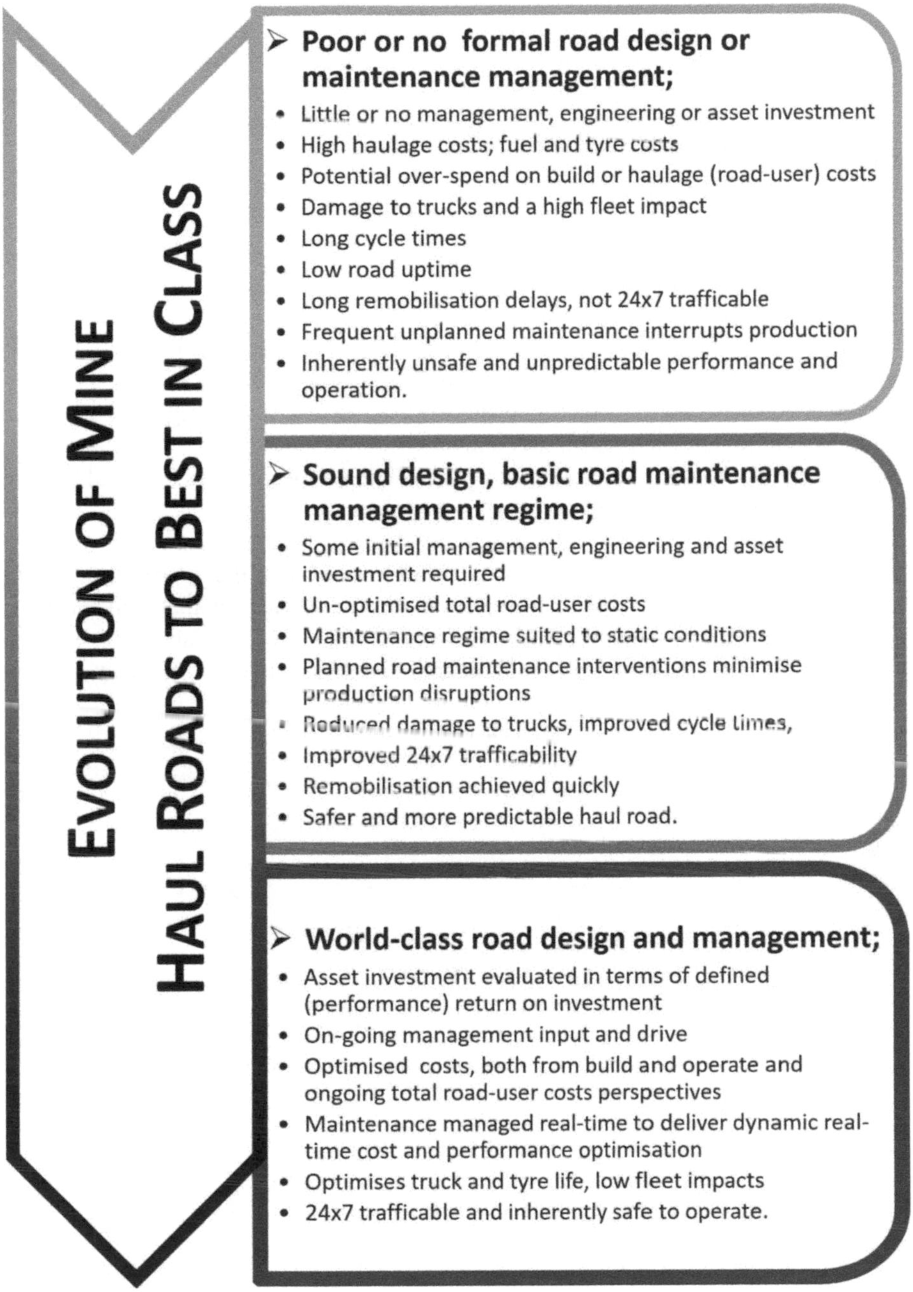

EVOLUTION OF MINE HAUL ROADS TO BEST IN CLASS

➢ **Poor or no formal road design or maintenance management;**
- Little or no management, engineering or asset investment
- High haulage costs; fuel and tyre costs
- Potential over-spend on build or haulage (road-user) costs
- Damage to trucks and a high fleet impact
- Long cycle times
- Low road uptime
- Long remobilisation delays, not 24x7 trafficable
- Frequent unplanned maintenance interrupts production
- Inherently unsafe and unpredictable performance and operation.

➢ **Sound design, basic road maintenance management regime;**
- Some initial management, engineering and asset investment required
- Un-optimised total road-user costs
- Maintenance regime suited to static conditions
- Planned road maintenance interventions minimise production disruptions
- Reduced damage to trucks, improved cycle times,
- Improved 24x7 trafficability
- Remobilisation achieved quickly
- Safer and more predictable haul road.

➢ **World-class road design and management;**
- Asset investment evaluated in terms of defined (performance) return on investment
- On-going management input and drive
- Optimised costs, both from build and operate and ongoing total road-user costs perspectives
- Maintenance managed real-time to deliver dynamic real-time cost and performance optimisation
- Optimises truck and tyre life, low fleet impacts
- 24x7 trafficable and inherently safe to operate.

Figure 1.3 Evolution in mine road design and management.

ii. under-expenditure on road design and construction, leading to premature failure; excessive truck operating costs; loss of productivity and, in the case of longer-term, high-volume roads, high contributory costs from rolling resistance effects. Under-designed roads are often maintenance intensive, to the extent that even well-built roads appear to perform poorly, due to maintenance being postponed on these roads to accommodate the intensive conservation requirements of the under-designed roads.

As tonnage increases and larger haul trucks are deployed, not only would the maintenance costs of existing roads of inadequate design increase, vehicle operating and maintenance costs also increase prohibitively. Formalising the approach to mine haul road design enables mining operations to benefit fully from the following:

i.   a safe, world-class road for all road-users;
ii. lowest vehicle operating costs, as a result of good haul road quality leading to faster cycle times and thus higher productivity and lower cost per ton;
iii. less stress on drive train, tyres, frame and suspension resulting in higher asset utilisation (less downtime) and component life; and
iv. optimum performance at minimum cost per ton hauled for the truck fleet and lower capital investment.

For these requirements to be fully and cost-effectively met, the approach to mine road design needs to consider and accommodate the range of haul trucks available to various types and sizes of mining operations. Figure 1.4 shows the range of available rear dump truck payload capacities (in metric tonnes) available from the major suppliers of Belaz, Caterpillar, Hitachi, Komatsu, Liebherr and Terex.

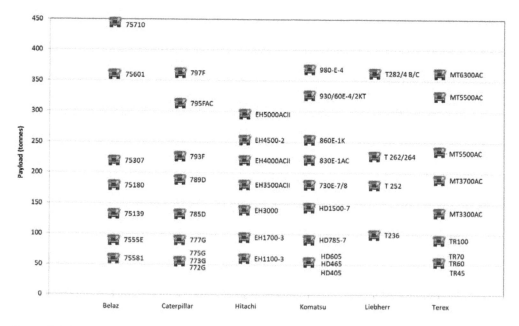

Figure 1.4 Range of payload capacities available for OTR rear dump mine trucks.

Following from the potential benefits, it is evident that there is considerable truth in the statement "haul roads can be profitable if properly designed but can cost you dearly if not carefully designed". However, direct financial benefits extend well beyond profitability and dividends, contributing also to the broader issues of sustainable development (SD). Contributions to a sustainable mining industry that can be derived from focusing on the mine haulage component of the mining cycle are critical and central to many SD reporting strategies. Typical SD reports encompass specific targets relating to energy and water consumption, greenhouse gas (GHG) emissions and safety and innovation. Improvements in haul road design and management can contribute positively to long-term sustainability solutions, in a relatively short-term implementation timeframe. Short-term value can clearly be derived from reduced energy, water and safety-hazards associated with mine haulage, whilst any costs associated with GHG emissions levied in the future also leverage longer-term benefits.

1.2 Characteristics of mine haul roads

Historically haul roads were unengineered, using the in-situ material as both the road structure and wearing course. Since the in-situ material did not necessarily have the support nor provide a safe and all-season use – especially in areas of deep weathering or regolith, the performance of these materials as a haul road were often unpredictable and unsatisfactory, both from a safety and economic perspective.

One of the first, and arguably most important, initiatives to formalise the approach to design and management of mine haul roads was the *USBM Information Circular 8758 – Design of Surface Mine Haulage Roads – A Manual,* by Kaufman and Ault (1977). The aim of this publication was to provide a broad manual of recommended practices that promote safer, more efficient haulage. The authors recognised that the development of surface mine haulage equipment had outstripped available (mine) road design technology, resulting in numerous accidents caused by road conditions that were beyond the vehicle's and driver's ability to negotiate safely.

The content of the USBM design guidelines was developed primarily in response to haulage accidents, but also included current (1970s) practice information from mining companies and equipment manufacturers. Content covered such aspects as road alignment (both vertical and horizontal), road cross-section, construction materials, surfacing materials, road width, cross-slope and berm design, together with traffic control and drainage provisions, as was suggested criteria for road and vehicle maintenance and for runaway vehicle safety provisions.

The USBM Guidelines adopted structural (layerworks) design procedures following the CBR thickness design procedure (Tannant and Regensburg, 2001; Thompson, 2011a; Yoder and Witczak, 1975, and discussed in Chapter 3 of this book), but based on the comparatively small haul trucks that were operated in the 1970s. Recent research by Thompson and Visser (1996) applied mechanistic analysis procedures which showed that the CBR method does not necessarily give the most effective road structural composition, especially as larger wheel loads, typical of today's ultra-class haul trucks, are now more commonly used.

Haul roads, particularly in surface mines, have generally been unpaved, which means that natural materials such as soil or gravel have been used as a wearing surface. The rationale for using untreated material, in addition to cost and performance considerations, was that, often, large rocks would spill onto the road and, on an unpaved wearing course, the material could be easily removed with motor-grader or front-end loader without permanent damage to the road. This policy has consequently been adopted by most surface mines.

However, with a dust-free road having become a requirement from a safety and environmental perspective, semi-permanent surfaces using a chemical additive have become a common solution in many mines, but with varying degrees of success. Although the surfaces created through these chemical treatment options are firm, any major spillage can still be removed by motor-grader or front-end loader without significant road damage – especially where repairable treatments are used. The USBM Guidelines also considered wearing course material selection, and more recent work by Thompson and Visser (2000a) has expanded material selection guidelines to capture characteristics of ideal wearing course materials, indicators of poorly performing materials and the effects on road rolling resistance and safety, as described in Chapter 4, together with the most recent learning in the use of the various chemical treatment options described previously.

In underground roads, concrete slabs are frequently used as a wearing surface in high traffic areas because of the presence of excess water and a highly abrasive environment through the action of tyres on the road surface. Inclines are, however, most often covered with locally available natural materials and similar functionality issues often arise when assessing the performance of these materials.

A cross-section of a typical haul road used on surface mine is shown in Figure 1.5 together with a description of the road structure terminology. Note that in civil engineering the term 'road pavement', which supports vehicles, is distinct from 'side walk', which pedestrians use.

In the case of surface mine haul roads, the road structure is provided to protect the in-situ material, which may be a weak material such as sand or clay, or a more resilient blasted rock or solid material. The thickness of the road structure or layerworks, also known as the sub-base and base, depends on the properties of the in-situ material, the quality of the

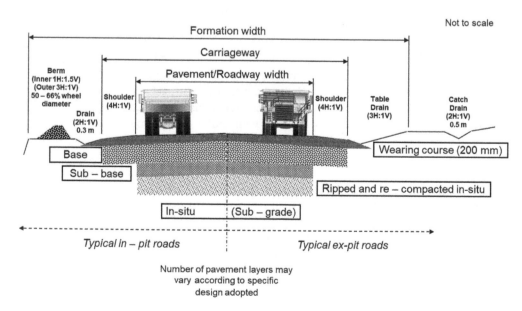

Figure 1.5 Cross-section of a surface mine haul road structure in-pit (LHS) and ex-pit (RHS) showing terminology of the main features incorporated in the design.

base material and the loads that need to be carried over the design life of the road. The base provides the majority of the structural support, whilst the wearing course or surfacing is the functional layer, providing a safe, economic and vehicle-friendly ride that can also be easily maintained.

On underground roads a base layer may be necessary to provide an even working platform as a result of overbreak in the mining process even though the tunnel may be in rock. Also, if drainage occurs along the rock/road interface, then the addition of a base layer will prevent ingress of moisture into the wearing course. Depending on the design, a two-layer weak cement stabilised layer and a higher strength and wear-resistant concrete slab may form the wearing surface and support for the wheel loads.

1.3 The provision of mine haul roads

The location and position of a mine haul road is dictated to a great extent by the mining method used and the geometry of both the mining area and the orebody. Mine planning software enables various haul road geometric options to be considered and the optimal layout selected, primarily from a lowest cost of provision perspective. Whilst these softwares often have default geometric design values embedded, it is nevertheless necessary to review the basic concepts of geometric design if any modifications are to be considered in the design of mine roads, either on the basis of economics or, more critically, from a safety perspective.

Geometry relates to the horizontal placement and curvature and the vertical alignment, namely vertical gradients and curvature in the transition between gradients. Since the geometry is relatively fixed, little can be done during operations to change the geometry, and this demonstrates the importance of the geometric design requirements and specifications being available at the planning stage. For new mine projects, this approach is quite straight-forward and defined through the planning process; most of the geometric design challenges that arise are most often associated with expansion projects to existing operations. Especially with the purchase of a larger truck fleet, not only would the existing pavement design be potentially inadequate from a performance (structural and functional) perspective, due to the constrained physical space in-pit, it would become difficult to accommodate the required changes to geometric design without having significant impacts in additional stripping ratio and costs. Therefore, the safety and operational benefits of modifying the primary haul roads to accommodate the larger truck types envisaged needs to be weighed against the costs associated with potentially increased stripping. This clearly becomes more problematic the deeper the pit is. Ex-pit roads, however, are less constrained (especially in the context of strip mining), but this is of little consolation in the many cases where the majority of the haul is in-pit, as is typical with many open-pit operations.

Once the geometric design requirements are determined, the remaining three stages in the provision of haul roads, namely the structural design, the functional design and the maintenance design, can be addressed. The design process is interconnected, in that the design requirement of each subsequent stage is informed, to an extent, by the previous stage. These stages are shown in Figure 1.6.

The structural design ensures that the road can support the anticipated loads, not only once or twice but many times during the anticipated design life of the road, which may be 3 months on a ramp or bench road to 20 years for the main haul road on the surface. Provided that the road structure has the desired support, this will ensure that the wearing course, which has special characteristics and provides functional performance, will offer the desired

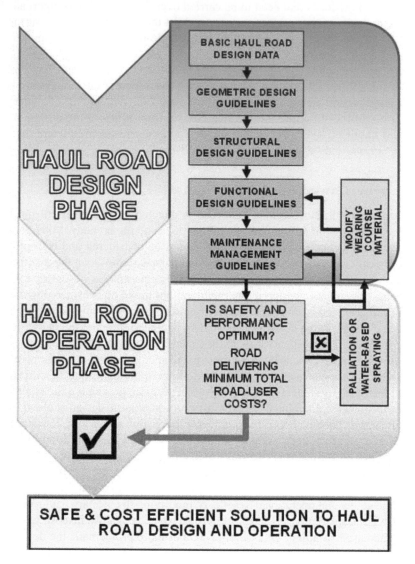

Figure 1.6 The interconnectivity between structural, functional and maintenance design.

performance and low rates of deterioration, such as caused by depressions, rutting, potholing, raveling, etc.

Every facility, irrespective as to how well it has been designed and constructed, requires maintenance to counteract the influence of use and environment. On haul roads, this has typically been by means of watering and motor-grader maintenance. A poorly designed and constructed haul road may require much more maintenance than a well-constructed road. As a rule of thumb, once a road has deteriorated, it takes 500% more time to fix it than it took to originally build. The better the roads are built, the slower the deterioration rate and the less maintenance will be required. The use of an appropriate road maintenance management strategy will generate significant cost savings by virtue of a better understanding of the relationship between wearing course material degeneration rates (manifest as increasing rolling resistance on the road) and its influence on both cost per ton hauled and the cost of road maintenance itself. These three aspects of haul road design and provision form the basis for the contents of this book, as ultimately the total cost of haul road construction and maintenance as well as vehicle operating costs will determine the economics and cost-effectivity of the operation.

1.4 The quality of mine haul roads

Mine haul roads are a unit of production, and the quality is related to minimising the total transportation cost, which consists of the sum of the vehicle operating costs and road provision and maintenance costs. Ideally mine management systems should be able to determine the vehicle operating costs in real time, but unfortunately this is not yet possible. Gleisner *et al.* (2017) presented a hypothetical simulated analysis of a mine route where the road quality, in terms of rolling resistance, was varied and the resultant cost per ton calculated using TALPAC® software (Haulage and loading simulator from Runge Pincock Minarco Limited) as shown in Figure 1.7. Truck size, transportation distance, mine production and average grade, among other parameters, were considered. The main input was a production target of 150.000 t per day and an analysis time horizon of five years. This shows the type of information that is desired in real time.

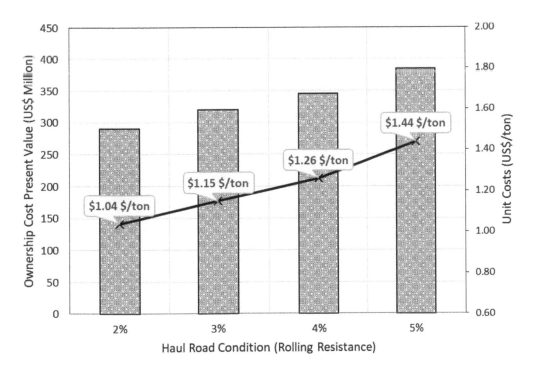

Figure 1.7 Sensitivity of haul road quality to transportation cost
Source: modified after Gleisner et al., 2017.

These benefits are similar to those provided by Thompson and Visser (2006a) who stated that for a fleet of 290 t payload, 498 t gross vehicle mass (GVM) rear dump trucks operating on a 4 km 10% incline, if the road rolling resistance is reduced from 6% to 2%, the capital cost of equipment necessary to move 30 million t per annum reduces by 18% whilst the truck operating costs reduce by 16%.

Although it is evident that real time costing is required to identify road quality, such costing is only available on a historical basis when quarterly or annual accounts are prepared. This does not permit the implementation of maintenance to improve road quality at the time it is necessary. A surrogate for such real time costing exists, namely rolling resistance as was described previously. Rolling resistance is the resistance to motion of a vehicle, and may be considered as an additional gradient as shown in Figure 1.8. Rolling resistance is an energy loss as a result of deformation of the tyres by rocks or road unevenness, sinkage of the tyre into the surface such as soft clay or loose material or deflection of the road structure under the wheel load, such as illustrated in Figure 1.8.

Rolling resistance may be measured by using coast down techniques, where a vehicle is permitted to coast with the gears disengaged and the deceleration is measured in both directions to eliminate the influence of gradient. This type of testing is not done routinely on mines as there are potential hazards with a truck travelling without the gears being engaged and most large mine haul trucks have electronic drive interlocks to prevent this type of operation (Thompson and Visser, 2006a). Another research method is using a single axle trailer with variable load and measuring the draw-bar force, which is related to rolling resistance.

Since the research focused methods for determining rolling resistance are not suitable for day-to-day measurements, Thompson and Visser (2003b) developed a visual inspection method which is based on the extent and severity of road defects that affect rolling resistance, several of which are illustrated in Figure 1.9. This procedure was developed on road sections where there was a range of defects and where the actual rolling resistance was measured by the coast-down method. This procedure will be discussed in detail in Chapter 5 where road management and inspections are considered.

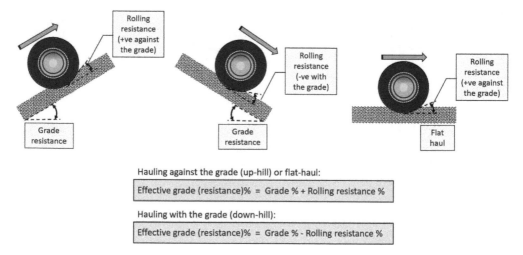

Figure 1.8 Rolling resistance is an additional gradient as a result of resistance to motion.

Figure 1.9 Examples of contributory road defects which increase rolling resistance, corrugations, rutting and tyre sinkage.

Figure 1.10(a) and (b) shows the effect of road condition on rolling resistance which was measured. These rolling resistance values may be compared with internationally benchmarked ideal rolling resistance of 2–2.5%. In most of the simulation packages for mine operations, a value of 3% rolling resistance is used.

(a) Rolling resistance 6%

(b) Rolling resistance 2.5%

Figure 1.10 Photographs of haul road condition and associated rolling resistance.

The consequence of assuming a low rolling resistance during the planning phase means that productivity is over-estimated and haulage costs are significantly underestimated if the actual rolling resistance is much higher. In instances where a ramp road has a gradient, which together with the 3% rolling resistance is at the limit of the capabilities of a loaded mine truck, the truck may not be able to exit the pit fully laden if the actual rolling resistance is greater than 3%, and this will severely affect productivity. This will be discussed in more detail in Chapter 4 where the interaction of functionality (wearing course selection) and rolling resistance are discussed.

As was pointed out in the beginning of this section, rolling resistance has important cost implications. In simulations of an electric drive rear dump truck (4.27kW engine power per tonne GVM) on a ramp road of 8–10% grade (against the load) and a basic rolling resistance of 2%, an additional 1% rolling resistance will reduce truck speed (and increase travel time) by 8–11%, whilst on a flatter surface road of 0–2% grade (against the load) and a basic rolling resistance of 2%, an additional 1% rolling resistance will reduce truck speed by between 18% and 23% when the vehicle is operating at full throttle. This much larger effect is due to the higher speed of the truck on these flat haul sections.

Figure 1.11 shows the effect of gradient and rolling resistance (i.e. effective grade) on typical mechanical and electric drive vehicles powered at full throttle with approximately similar kW engine/ton GVM. The electric drive vehicle maintains a slighter higher speed at 2–3% effective grade (i.e. a flat haul with 2–3% rolling resistance). However, at typical ramp gradients of 8–10%, both types of vehicles maintain similar, but significantly lower speeds. The gradient of the speed-effective grade curves suggests that these vehicles are most sensitive to increases in rolling resistance associated with flat-haul segments of the network. With the correct selection of wearing course material, coupled with an appropriate maintenance strategy, the rate of deterioration (or increase in rolling resistance) can be minimised, offering a road that performs to design specifications for longer, on which trucks can maintain higher average hauling speeds.

Figure 1.12 shows that rolling resistance, when acting against the load, will reduce truck speed and productivity, cause excessive fuel burn (one of the most significant costs associated with truck haulage operations) and reduce tyre and component life and increase maintenance

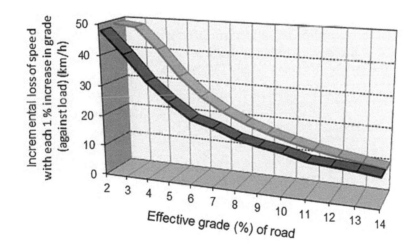

■ Mechanical Drive	▢ Electric Drive
180 t capacity, 317 t GVM rear dump truck (mechanical drive) with 1416 kW (1336 kW @flywheel) engine power, equivalent to 4.21 kW/t GVM	194 t capacity, 324 t GVM rear dump truck (electric drive) with 1492 kW (1389 kW @flywheel) engine power, equivalent to 4.27 kW/t GVM

Figure 1.11 Effect of gradient and rolling resistance on vehicle speed.

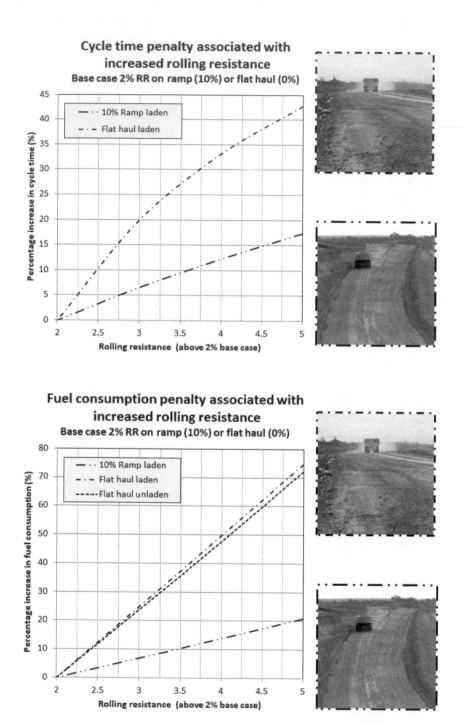

Figure 1.12 Rolling resistance impacts on cycle time and fuel consumption for a selected haulage cycle comprising ramp and flat-haul segment.

costs. Using cycle time estimates from OEM speed-rimpull-gradeability data and a typical haul cycle combining ramp and flat-haul segments;

- Each 1% increase in rolling resistance above 2% base case increases cycle times by 20% on flat roads and 8% on ramp roads.

  - 2km ramp cycle from 13'06"→15'24" (laden and return)
  - 2km flat haul cycle from 5'12" →7'54" (laden and return)

Thus the haulage component of the cycle time increases by approximately 21%.

For fuel consumption estimates, generated from truck fuel consumption data and speed estimation models, for the same haul cycle described previously:

- Each 1% increase in rolling resistance above 2% base case increases fuel consumption by 25% on flat roads and 5% on ramp roads.

  - Ramp cycle fuel consumption from 85 →103 l (laden + return)
  - Flat cycle fuel consumption from 35 →60 l (laden + return)

Thus causing a total fuel cost increase of approximately 28%.

Although the quantum of benefit from improved haul road design, operation and management will vary considerably both with the type of mining method and especially traffic volumes and tonnages hauled, the value-add is universal; any operation will derive benefit from the application of improved haul road engineering. The focus of this book then is to introduce the key concepts and principles of mine haul road design, such as philosophy of provision, geometric alignment, structural, functional (wearing course) and maintenance designs and, importantly, their practical application in mining. By applying the theory and practice described, mining companies can ensure they leverage the best possible return on investments made in the design, construction and operation of a haul road network, to the direct benefit of the organisation, its employees and shareholders.

Chapter 2

Concepts for geometrical design

2.1 Geometric design and safer haul roads

Accident prevention strategies evolve from sound fundamental engineering design aspects, coupled with enforcement and education, together with human error behavioural analysis, to better understand and control or eliminate hazards. In this chapter, the key considerations from a geometric, or horizontal and vertical alignment design perspective, are considered. However, as alluded to in Chapter 1, with both a new or existing road, any modifications to accepted geometric design guidelines established at a mine-site need careful consideration since geometric design exerts significant influence over cost and profitability, in terms of pit space required or ore sterilised and, more critically, from a safety perspective. As an introduction to geometric design, the role of a systems approach to identifying and addressing safety-related design issues is first introduced, as a basis for the further development of geometric design.

In the context of mine roads, requirements may be as diverse as these:

- In- and ex-pit (light- and heavy-vehicle) haul roads
- Ex-pit access roads
- Infrastructure service roads

Although each of these roads requires a different approach to design, there is commonality in terms of a broad safe systems approach, such as one outlined by the Australian Transport Council (2006), in which the identification, removal or amelioration of road elements which may contribute to hazards and accidents is a key component, irrespective of the type of road being designed. A safe system acknowledges that humans are fallible, error is inevitable and that, when it does occur, the (mine haul) road system makes allowance for these errors so as to minimise the level of severity associated with the risk.

A safe systems approach to mine road design (modified after Australian Transport Council, 2006 and Vagaja, 2010) requires the following:

- Designing, constructing and maintaining a road system (roads, vehicles and operating requirements) so that forces on the human body generated in accidents are generally less than those resulting in fatal or debilitating injury.
- Improving roads and adjoining areas to reduce the risk of accidents and to minimise hazards: designing 'forgiving' haul roads and roadsides which allow a margin of recovery from error.
- Managing speed, and taking into account the speed-related risks on different parts of the road system.

Using this approach, and recognising the three distinct mine road systems, in- and ex-pit haul roads (as discussed in this book) and also ex-pit access roads and infrastructure service roads, the key elements of the safe systems approach are shown in Figure 2.1.

2.1.1 Integrating design methodology with safety audit systems

In the mining environment and specifically with in- and ex-pit haul roads, safer vehicles are addressed through local and international earthmoving equipment standards. Safer roads are addressed both through sound engineering and integrated design methodologies, coupled with the safety auditing of each design stage. (For access and infrastructure service road design for light and conventional on-road vehicles, existing design guidelines for unsealed 'conventional' roads would typically apply.)

The integrated road design methodology shown in Chapter 1 considers each design element as part of the overall 'design' process. A safe-systems approach should link the overall design process to broader safety considerations whereby other road-safety elements are considered in the context of the network of roads. Safety audit systems (such as the AustRoads (2009b) method) have been used on public road networks for a number of years and more

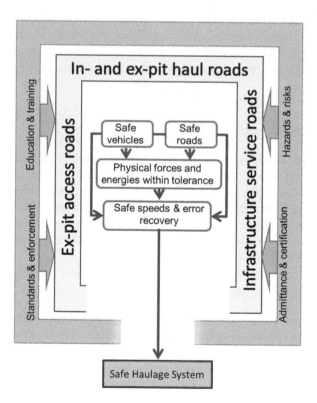

Figure 2.1 Safe systems framework for the provision of mine roads.

Source: modified after Australian Transport Council, 2006.

recently the adaptations developed by Vagaja (2010) have been applied in mining operations where typically the following mine-site road-safety auditing elements are included:

- Evaluation of current safety system documentation, standards and processes
- Vehicle and/or road-user interactions
- Speed management
- Geometric layout including potential areas of traffic conflict
- Signage, delineation and lighting
- Parking arrangements, including 'go-line' design and HV, LV and pedestrian separation

To prevent an accident or reduce the severity of its consequences, a transport system (i.e. not just the road 'design' but all the associated operational features and vehicles) should also explicitly consider human error. In this way a haul road design process can be audited to identify interactive effects and thus potential error escalation is minimised. Figure 2.2 shows how the approach to mine (heavy vehicle) road design can be combined with the generic road-safety audit systems approach. The combination of rigorous design and systems interaction auditing enables critical deficiencies in design and operation to be flagged and addressed throughout not just in the design (concept, draft and feasibility) stages, but also in the operational phases of the haul roads.

2.1.2 Analysis of mine haul road and haulage incidents

Mine haulage incident records should be kept as part of the safe systems approach described earlier. These records can be used to identify the major contributory factors that led to these incidents as a basis for a more comprehensive and informed approach to mine haul road design.

Several haulage safety or incident analysis studies have been undertaken, both in the United States (Aldinger et al., 1995; Hunting and Weeks, 1993; Kecojevic and Radomsky, 2004; Randolph and Boldt,1996; United States Bureau of Mines (USBM), 1981) and elsewhere, for example by the South African Mine Health and Safety Council (MHSC, 2001) and the Minerals Council of Australia (MCA, 2006). From an analysis of the principal substandard surface mine haul road design factors which were most frequently encountered in incident reports, these studies led to the identification of several key problem areas which guide and inform the design process.

In the 2001 South African Mine Health and Safety Council study (MHSC, 2001) complete incident and accident reports were evaluated to better determine incident causation and Figure 2.3 presents the various percentages of agencies implicated in those incidents (after Fourie et al., 1998). Results generally echo similar studies in that of the total transport incidents analysed and categorised, 47% could be directly attributed to road design and haulage operations. Of these attributable incidents, 40% were associated with sub-standard road design factors, geometric and functional components predominating as the agencies implicated, with maintenance and structural design deficiencies exhibiting less influence.

It was also accepted that in the majority of incident reports analysed, scant attention or recognition was given to basic road design components, even where the deficient condition which led directly to the accident was clearly stated. It was shown in this work how the 'formality' of the design process related to incidence rates – the more 'formal' a haul road design process is, the less is the attributable accident rate, a result which, in part, informed the integrated haul road design process referred to in Chapter 1.

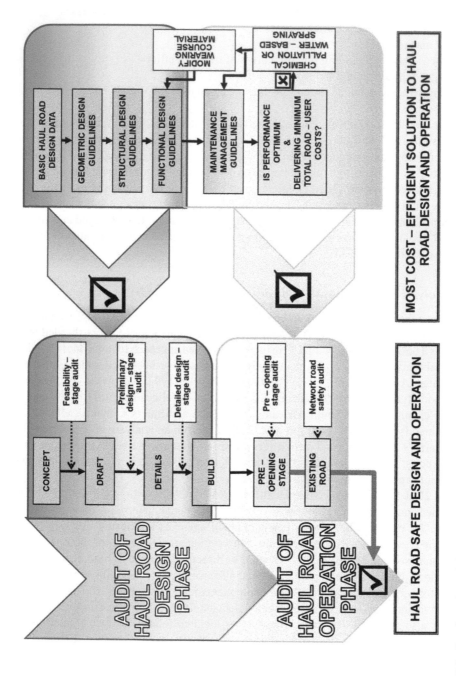

Figure 2.2 Combined road design and design safety audit methodology.

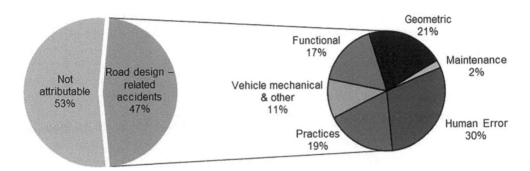

Figure 2.3 Relationship between road design and attributable accident rates (MHSC, 2001)

For the proportion of accidents related to road geometry, the principal deficient road geometric design considerations were found most often to be related to the following:

- Intersection layout:
  - Poor intersection layout or incorrect or inappropriate signage. Poor visibility of or from junction. Non-uniformity of traffic controls
- Safety berms:
  - No safety berms where road runs on an embankment (fill area) or berms too small. No berm maintenance. Vehicles which lost control on these sections ran off the road
- Road shoulders:
  - Collisions with vehicles (breakdown, etc.) parked on roadside, no shoulder or road too narrow. Poor demarcation of parked equipment. Poor or temporarily obstructed sight lines
- Run-aways/brakes:
  - Accidents due to brake failures whilst hauling laden down-grade or vehicle run-aways down-grade. Excessive gradients (>10%)

2.2 Integrating design methodology with mining plan

Hustrulid and Kuchta (2006) presented the fundamental concepts guiding the location of a haul road ramp from bottom to top of an open-pit. Key considerations are these:

- Longitudinal grade
- The difference in height between the mining benches
- Ramp location – inside or outside the wall
- Volumetric and stripping ratio implications of ramp location
- Other associated mining impacts from the road geometry

Figure 2.4 presents the principles to develop the layout of a ramp haul road, using mining software, going from the bottom to the top of a pit, inside the wall. This example shows a 10% longitudinal grade road with 20 m width coming out of level 300 m to the exit at level 340 m.

It must be noted that the placement of a haul road is an interactive exercise in which the impacts of various location options are assessed on the mine economics. The starting point and the ending point of a road are dependent on the route the ramp will take in between those two points. The latest versions of mining software design packages (such as DATAMINE Studio OP® or MineSight ®) provide tools that make the life of the mine planner easier when it comes to include the road projection in an integrated mine design. When the haul road allocation is made within the pit design it is normally done from the bottom to the top of the pit, considering the optimal pit shell provided by the optimisation software (e.g. NPV Scheduler® or Whittle®). The definition of the starting point (elevation) would be the location in the lowest level of the pit that allows enough room for the mining equipment (trucks, excavators, dozer, etc.) to maneuver and operate. Inside that area, the coordinates could be anywhere in the boundary polygon that represents the toe line of the lowest slope. Once location is established, either a clockwise or anti-clockwise direction can be applied to reach the end point of the in-pit ramp. The end point of the ramp will either be an intersection with existing surface haul roads, a waste dump (if waste material is being hauled) or the processing plant (if ore is being transported). Low grade ore can be stockpiled in different locations too and, consequently, access to those spots could also be planned. These choices define the alignment of the axis along which the road will be projected.

Conceptually, the interactive analysis aims to combine reductions in transportation distances with reductions in load elevations, whilst at the same time minimising the stripping ratio and potential ore sterilisation (either high or low grade) due to haul road locations in or above blocks to be mined. These considerations are the main aspects to take into account

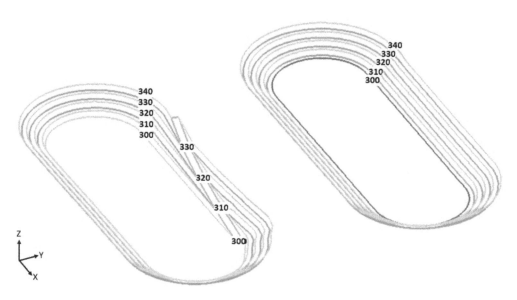

Figure 2.4 The geometrical approach of building a road inside the pit wall.

when determining the most economically attractive configuration in terms of value (measured by the NPV, for example). However, the decision itself is subject to review as in most operations, a better estimate of the orebody geometry and ore grades become available when short-term drilling data is acquired. This improved knowledge then affects the previous long-term pit design, and consideration of the new mine model will redirect the pushback design and consequently the placement of the roads.

The ramp road itself can be either spiral around the inner perimeter of the pit, or incorporate switchbacks to keep the road on one side of the pit, or a combination of the two alternatives. Switchbacks are sometimes necessary in steeper pits, to reach some inaccessible levels or to shorten haul distances, but in general the preferred option for ramp design would be for spiral roads. In the switchbacks, the space for two vehicles is normally tighter and, as a result, they tend to slow the traffic or even require the empty vehicle to stop and obtain clearance prior to proceeding through the switchback. Due to the tighter radius of curvature, the vehicle tyres at this point are exposed to higher wear and the wearing course itself is normally subject to high maintenance frequencies due to the effect of high horizontal shear forces. From the safety point of view, it is also problematic when the visibility of the trucks is limited. Ideally, from a planning perspective, some strategies should be used to ameliorate these problems, such as locating the switchback on a flat (as opposed to grade) areas, reducing the longitudinal grade and increasing the width of the road.

2.2.1 Including haul roads in the ultimate pit design

From the strategic mine planning perspective, the road design is normally done in the final pit design stage to ultimately define the ore reserves. Whilst some of those designs are impractical at various stages of the mine life, they exist just as a design to comply with standard requirements of reserves statement reporting.

In the example shown in Figure 2.5, contours from the final pit shell show the pit bottom at an elevation of 36 m; the uppermost level seen is 236 m. Considering a ramp with 10% longitudinal grade from the pit bottom to the highest level, a 2000 m-long ramp will be necessary to overcome the 200 m difference between the levels. (Although this does not mean that this is the total length of a single road, but rather the expectation during the mine life of the in-pit road network required to access all the benches). Those contours will now be used to control the pit expansion and determine the pit design adherence between the optimal pit shell (mathematical surface from the optimisation algorithm) and the final pit design including the haul roads.

Figure 2.5(a) shows a plan view of the pit (with contours and labels representing the elevation in meters), Figure 2.5(b) shows the pit bottom in detail, Figure 2.5(c) is a lateral view of the contours and Figure 2.5(d) is a perspective view of the pit contours that will be used as a reference to place and build the ramp road inside the ultimate pit design. The arrows represent the end point of the in-pit road to access the processing plant and the end point to access the waste dump. The contour of level 36 m has an area of 1637 m^2, and the contour of level 44 m has an area of 4384 m^2. If the area available is too small to accommodate all mining activities, the pit bottom could be established at the contour level of 44 m and a temporary (narrower and/or steeper) road can be built to reach level 36 m. In this road placement example, the starting point will come from the pit bottom at level 36 m and reach the endpoint at level 162 m. This will be considered as the road to take the ore from the pit bottom to the processing plant to be included in the final pit design, as shown in Figure 2.6.

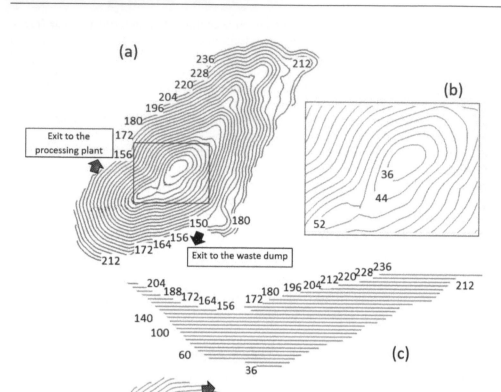

Figure 2.5 Contours from an ultimate pit optimisation.

In Figure 2.7, a perspective view of the pit design is presented showing the main haul road to transport the ore (orange), the pit design (red in the slope faces and green in the berms) and the external topography (grey).

2.2.2 Integrating roads through the mine schedule

As discussed in the previous section, the haul road should be included within the final pit design; if no significant orebody changes and/or new mineralised zones are discovered during the mine life, this design should be pursued throughout the mine life. This means that the design of the short- and medium-term roads should progress to connect to the long-term pit design. The connection of a short-term design with the long-term design determines which parts of the road are considered permanent and which parts of the roads are temporary, together with the

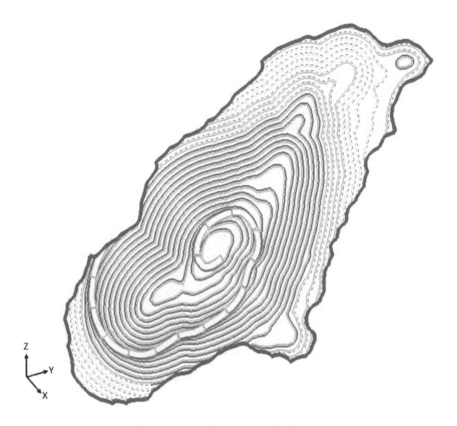

Figure 2.6 Ore haul road included in the ultimate pit design.

durations of each over the mine life. It is intuitive that the mine will expand (horizontally and vertically) over time. With that constant change and the dynamism of the mine, it must be clear to the mine planning engineer what to do when projecting a new pushback for a further mine expansion. From the haul roads point of view, the analysis should consider aspects like these:

* Which part of the road should be kept?
* Which part of the road should be extended?
* Which part should be replaced and rebuilt?

Figure 2.8 shows an example schedule for the deposit analysed, the colours represent the years of the mine sequence. Each set of blocks generate a pit shell that should be analysed individually to make the operational design for each phase (year) according the legend.

The individual pit outlines for each year are shown in a plan view in Figure 2.9. The coloured polygons representing the different years; year 1 to year 4 shows that the outlines are expanding to the south until it gets to the final outline in year 5 (red line). From year 5 onwards the pit expands essentially in depth and from this point onwards the in-pit roads should follow the deepening of the pit to allow the operation to move the ore to the processing plant (exit west) and in parallel the waste to the waste dump (exit southeast).

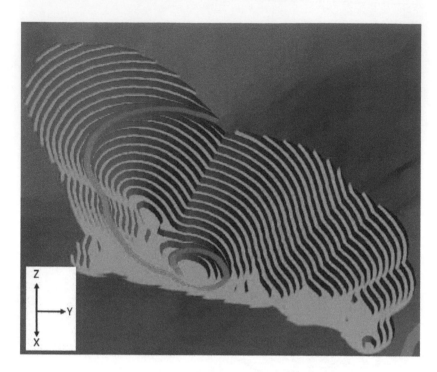

Figure 2.7 Perspective view of the pit design with the main in pit road.

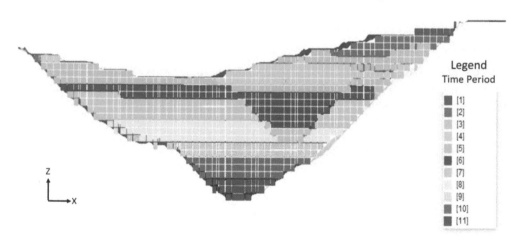

Figure 2.8 Cross section of the deposit showing the mathematical schedule.

In Figures 2.10 and 2.11, the evolution of the in-pit haul roads is shown for several phases (years) of the mining sequence. To demonstrate the evolution of the mine and the haul roads development, Figure 2.10 shows the pit design for year 1 with the final pit outline (year 11) shown in red. The haul roads for year 1 (represented in orange) are limited

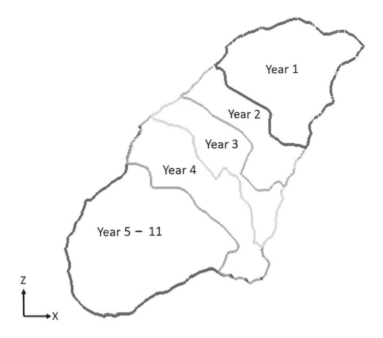

Figure 2.9 Pit outlines according the annual mine sequencing.

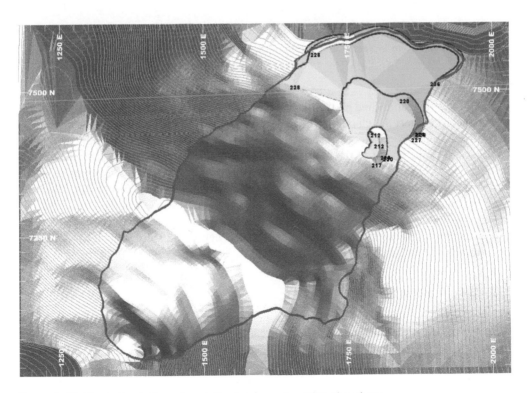

Figure 2.10 Pit design in Year 1 and the haul roads within the phase.

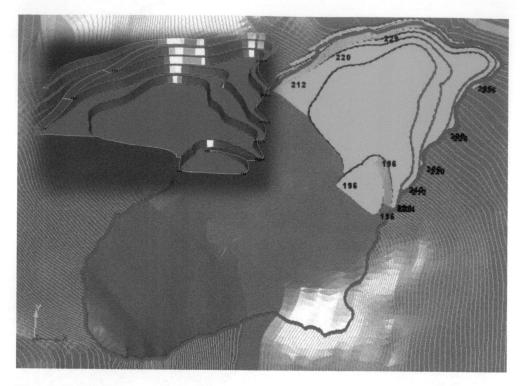

Figure 2.11 Pit design in Year 3 and the haul roads within the phase.

as the pit is still shallow and all the roads are considered temporary (Category III) or semi-permanent (Category II) since, for the most part, the pit is going to deepen vertically and expand horizontally and those roads will have to be repositioned in the new pushback. In Figure 2.11 the layout of the roads in year 3 are shown, and it is clear that none of the original (year 1) roads still exist, thus new roads were constructed for this iteration of the mine schedule. Likewise, as mining progresses, the network of roads can be identified and categorised according to life and traffic volume, thus, in the road design phase, a specific design methodology to meet these individual requirements can be applied, with consequent savings in both capital and operating expenditure when roads are designed for a specific and defined life and purpose.

2.3 Geometric design process

The geometric layout of a mine haul road is dictated to a great extent by the mining method used and the geometry of both the mining area and the orebody. The road layout – or alignment, both horizontally and vertically – is generally the starting point of the geometric design. Practically, it is often necessary to compromise between an ideal layout and what mining geometry and economics will allow, prioritising at all times any safety-related requirements, flagged from any of the design-stage audits as discussed earlier (Section 2.1.1). Any departure from the ideal geometric specifications will result in reductions of both road and transport equipment performance and, critically, expose operators to an inherently unsafe system. Considerable data already exists pertaining good engineering practice in geometric design, and many local standards, specifically developed for the local operating environment could apply. However, generic concepts are used as the basis of the design criteria developed here. The process of geometric design begins with a simple objective of connecting two points, and this objective is improved incrementally as the geometric specifications are applied and met. The steps are shown in Figure 2.12. Note that although the flow diagram is focused on a road, drainage of a road section has to link to the drainage master plan, as removal and safe storage (sumps) are integral to the mining plan.

Broadly speaking, safety and good engineering practice require haul road alignment to be designed to suit all vehicle types using the road, operating within the safe performance envelope of the vehicle (85% of maximum design vehicle speed as an upper design speed),

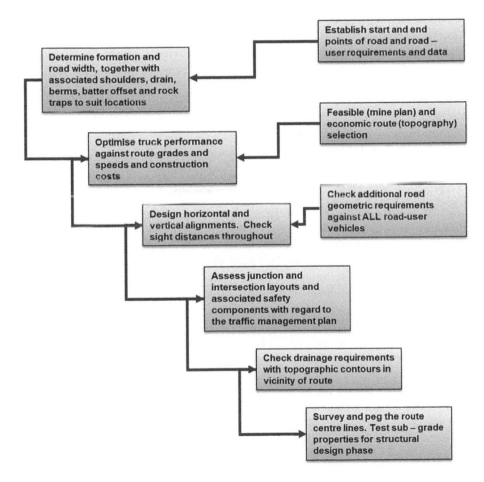

Figure 2.12 Simplified flowchart for the geometric design process.

or at the speed limit applied as dictated by the design itself. Ideally, geometric layout should allow the vehicles to operate up to the design speed, but since the same road is used for laden and unladen haulage, there is often the need to minimise laden travel times through appropriate geometric alignment, whilst accepting compromise (generally in the form of speed limits) on the unladen return haul. Critically, throughout the design process, reference is made to the local mine traffic management plan or traffic rules, including priority rules, etc. since this information would also inform the design procedures, alignment and road layout decisions.

Many of these additional considerations are dealt with in various Codes of Practice, for example: Worksafe Australia Roads & Other Vehicle Operating Areas Draft Code of Practice, and DMP (WA) Mobile Equipment on Mines High Impact Function (HIF) Audit, which includes considerations related to the following:

- Traffic management plans and traffic rules
- Mine access roads
- Road standards
- Separation and segregation of vehicles and pedestrians
- Restricted access exclusion zones
- Traffic movement around buildings, structures and service corridors
- Communications
- Lighting
- Traffic control signage
- Intersections
- Parking areas
- Road construction and maintenance

2.3.1 Stopping and sight distances considerations

Stopping distance requirements are a critical component of the geometric design process. Together with sight distance, they have a significant impact on the operational safety of a road. The truck manufacturer should ideally always confirm the distances required to bring a particular vehicle to a stop, following ISO 3450:1996 standards. This ISO standard, which specifies braking systems' performance requirements and test procedures for earth-moving machinery and rubber-tyred machines, is often used as a design standard by equipment manufacturers to enable uniform assessment of the braking capability of earth-moving machinery operating on work sites or public roads.

The ISO standard gives typically 114 m stopping distance at 10% downgrade from 50 km/h and 73 m from 40 km/h. Whilst this satisfies most mine ramp road designs where rear-dump trucks are used, care should be taken when using the ISO approach for articulated dump trucks (ADT). Steeper ramps are often used where ADTs are employed, since they commonly have better hill climbing ability. With a ramp steeper than 10%, the ISO stopping distance would not necessarily be met and, of course, what stopping distance is achieved in practice depends also on the coefficient of longitudinal friction supply (or skid resistance) of the road surface-wheel interface. Even if the braking system is capable of meeting or exceeding ISO 3450 requirements that does not imply that a vehicle will come to a standstill within that distance, as there are other factors in stopping distance calculations to consider.

In general, and including driver reaction times and importantly, brake system activation times to full braking effort, practical retard unassisted (emergency) braking distances can be determined from this equation:

$$d_b = \frac{1}{2}gt_{p-r}^2 \sin\theta + v_o t_{p-r} + \left(\frac{\left(gt_{p-r}\sin\theta + v_o\right)^2}{2g\left(U_{min} - \sin\theta\right)} \right) \qquad \text{Equation 2.1}$$

Where
d_b = Stopping distance (m)
g = Acceleration due to gravity (m/s²)
t_{p-r} = Driver reaction AND brake activation time (s)
θ = Grade of road (degrees) positive (+ve) downgrade
U_{min} = Coefficient of longitudinal deceleration (friction supply)
v_o = Initial vehicle speed (m/s)

A further simplification of Equation 2.1 yields this:

$$d_b = v_o t_{p-r} + \frac{v_o^2 - v_f^2}{2g(U_{min}\ BE \cos\theta - \sin\theta)} \qquad \text{Equation 2.2}$$

Where
v_f = Final vehicle speed (m/s)
BE = Braking efficiency (generally close to 100%)

Or, in alternative units, it forms:

$$d_b = \frac{V_o t_{p-r}}{3.6} + \left(\frac{\left(V_o\right)^2}{254\left(U_{min} - 0.01GR\right)} \right) \qquad \text{Equation 2.3}$$

Where
V_o = Initial vehicle speed (km/h)
GR = Longitudinal grade (%), -ve upgrade, +ve downgrade.

A reliable first estimate for stopping distance is based on 'ideal' braking and vehicle conditions (dry road, good skid resistance, tyres in good condition and at recommended pressure, etc.), with U_{min}, the coefficient of longitudinal deceleration (a component of friction 'supply' in response to the 'demand' imposed by the decelerating haul truck), being generally taken as 0.25 (under certain favourable conditions only). When conditions under braking vary (wet roads, poor and slippery wearing course, spillage, sub-standard tyres, etc.), a greater stopping distance would need to be considered. U_{min}, being <0.25 (wet, soft, muddy, rutted road surface), but can be significantly lower in certain circumstances, e.g. ice on road, high clay content in wearing course, etc.

Figure 2.13 shows how the coefficient of longitudinal deceleration (friction supply) and driver and brake system activation times affect stopping distance, compared to the ISO 3450 standard. Note should be taken of the combined effect of changes to U_{min}, the coefficient of longitudinal deceleration and also the driver's and the vehicle reaction and brake activation

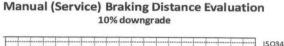

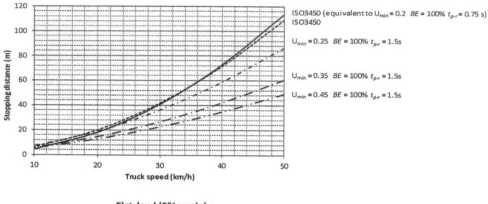

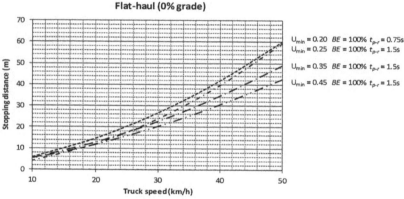

Figure 2.13 Braking distance evaluation at 10% downgrade, with various longitudinal tyre-road-surface friction and driver reaction times, compared to ISO3450.

time t_{p-r}. The solid line represents the ISO minimum stopping distance requirements of the braking system, which could easily be exceeded when conditions (U_{min}, t_{p-r} or braking system efficiency or brake fade) vary.

The coefficient of longitudinal deceleration (friction supply) classifications (normal and warning levels 1–3) are shown in Figure 2.14 following the original development and testing of RSTS (Road Safety Training Services – Tulloch/Stoker model). This model provides guidance on the operational warrants necessary when the measured (using a light vehicle – LV) coefficient of longitudinal deceleration (friction supply) falls below a threshold of 0.45, which is based on the equivalent (lower) levels of friction supply that would be mobilised for a haul truck under the same operating conditions.

Tulloch and Stockers's guidelines indicate that when LV-measured friction is maintained above 0.45, vehicle operation is safe (at least in terms of skid resistance and stopping distance requirements). Below that minimum threshold, three levels of warning are recognised: caution, potential hazard and hazardous for haulage operations. When LV-measured friction supply falls below 0.25, haulage should be suspended until the road surface dries back to at least

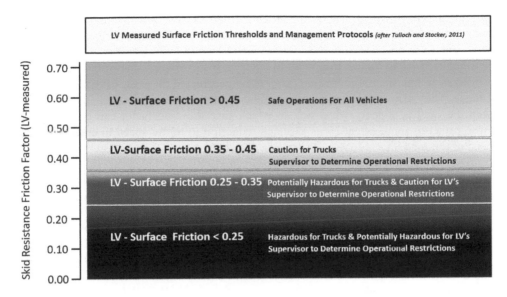

Figure 2.14 Haul road friction supply classifications (normal and warning levels 1–3).

Source: after Tulloch and Stocker, 2011.

minimum friction supply value of 0.25. This would represent the 'worst-case' scenario and stopping distances should be adopted based on this minimum value of LV-measured friction supply (assuming haulage is suspended when friction falls below this threshold).

Typically, Australian site experience suggests that between 0.4 and 0.8 mm of rainfall could induce changes in friction supply to below the desired threshold limit of 0.25. However, specification of a hard, aggressive larger size fraction in the wearing course could extend operability of the road under wet weather conditions, as will be discussed in Chapter 4.

The actual stopping distance value adopted for design purposes should be based both on the worst-case friction supply and on the vehicle with the longest stopping distance requirement. The degree to which this distance will have to be increased will depend on the type of wearing course material, moisture content, climatic conditions, type of tyre, inflation pressures, load and vehicle speed also. Friction supply cannot be continuously measured by a truck driver in-cab – only subjectively estimated from the 'look and feel' of the road. The safety issue is then if friction demand during braking is, for example, 0.25 and supply is 0.40, then there is a significant safety margin. If, however, the supply falls to 0.3, then there is only a 0.05 margin of error; if steering is simultaneously required, then this demands additional friction supply, so under these conditions there is little friction supply remaining to provide steering control.

2.3.2 Measurement of friction supply on mine haul roads

Skid resistance (or friction supply) is the force developed when a non-rotating tyre slides along the haul road surface. Friction involves two components; the tyres and the road surface itself, both of which are variable with wetness, speed and tyre type and condition, etc.

Due to the highly variable nature of these parameters, there is little information regarding typical skid resistance values for unsealed road surfaces for dry and wet states and fewer still referring to mine haul roads. Some estimates of friction factors for unpaved roads have been published by Fricke (1990) and Lea and Jones (2007). The latter authors identified a range of values from 0.40 to 0.85, with the lower value of 0.4 being recommended as an interim minimum drag factor (i.e. uncorrected for grade of road) for stopping distance. They also note the minimum value does not, however, account for the situation when the road is slippery when wet, in which case the drag factor can be significantly lower, offset to some extent by an assumed reduction in traffic speed to under the design speed of the road.

Limited South African research (Paige-Green, 1990) concluded that skid resistance in a dry state depends mainly on the mass fraction of the wearing course passing 26.5 mm, a measure of plasticity and fines content. Paige-Green's model for skid resistance is shown in Equation 2.4 as:

$$MD = 7.69 - Rain(1 - 0.1P_{0.075}) \times 10^{-3} - 0.018P_{26.5} - 0.004PF + 1.08DR \qquad \text{Equation 2.4}$$

Where
MD = Maximum deceleration (m/s^2)
$Rain$ = Mean annual rainfall (mm)
$P_{0.075}$ = Percent passing 0.075 mm sieve
$P_{26.5}$ = Percent passing 26.5 mm sieve
PF = Plastic factor (Plastic Limit × $P_{0.075}$)
DR = Dust ratio ($P_{0.075}/P_{0.425}$)
$P_{0.425}$ = Percent passing 0.425 mm sieve

The presence of water (as a mean annual rainfall, not an application rate per unit area) reduces skid resistance in proportion to the amount of fines <75 μm. However, the model does not account for the more sudden changes in skid resistance encountered on the application of water or rainfall measured per unit area. Decelerations are not converted to drag factors, and no corrections are applied for grade. However, the sections were nominally flat and tested in both directions, which would eliminate the need for grade corrections.

Technical Recommendations for Highways No. 20 (Committee of State Road Authorities, 1990) adopts Paige-Green's work and notes that, in dry weather, unpaved roads may become slippery if an excess of loose fine gravel (between 2 and 7 mm in diameter) accumulates on the road surface through raveling under traffic or poor blading practices. This layer behaves like a layer of ball bearings and the skid resistance is reduced practically to zero. This is especially a problem on corners.

A reliable assessment of road surface friction 'supply' (in response to the 'demand' imposed by haul truck braking) is based on the coefficient of longitudinal deceleration measured either with a haul truck itself, or, more readily, through the use of LV-based measurement and determination of either

i.   the equivalent friction supply available to a haul truck or
ii. through truck operational warrants that should be applied when the LV-measured friction supply falls below a threshold value (e.g. as shown in the Tulloch/Stocker model in Fig. 2.14).

Since the coefficient of friction at the tyre-road interface changes due to the presence of water, the testing methodology must incorporate an evaluation of the amount of water applied to the road, either as rainfall, or for testing purposes, using a water-truck spray. Water-truck delivery rates can be easily determined by collecting spray-water in foil trays located at set distances from the truck centre, for a number of trucks passes at set speeds (typically representing ramp and flat-haul watering speeds, typically lock-up second and fourth respectively). As each spray head is run, a profile of the delivery rate and coverage can be determined in mm/m², as further described by Tulloch and Stocker (2011).

Figure 2.15 shows a typical set-up for a water-truck calibration trials, whilst Figure 2.16 shows a coverage pattern determined with this methodology and Figure 2.17 the new coverage pattern obtained by recalibrating the spray heads. It can be seen how a much more even distribution pattern is achieved, which is better aligned to a good operating practice guideline of approximately 0.2 l/m² (equivalent to 0.2 mm/m²) water delivered to the road surface per spray pass.

Once the water-truck spray delivery rates are calibrated, the water-truck can then be used to assess the extent to which the application of water to a haul road wearing course contributes to a reduction in the coefficient of friction of the road surface.

The coefficient of friction is most easily measured by braking a vehicle from a given speed and measuring either the drag factor (a combination of coefficient of friction and grade of the road), or coefficient of friction (where grade is applied to adjust the coefficient of friction). In either case, the vehicle must brake and wheels must lock up such that an average value of friction, measured as average longitudinal deceleration, is found. Numerous commercial devices are available to assist with the measurement and recording process, such as the Vericom VC4000DAQ shown in Figure 2.18. Alternatively, many smart phones are now equipped with accelerometers which, when the phone is mounted and oriented correctly, can be used as a substitute, running a suitable application (e.g. SMYK from Styrax Tools, CDV Czech Department of Transport, or ADE – Australian Diversified Engineering).

Figure 2.15 Collector tray layout for simple water-cart delivery rate calibration calculation (with red trays indicating centreline of the truck path).

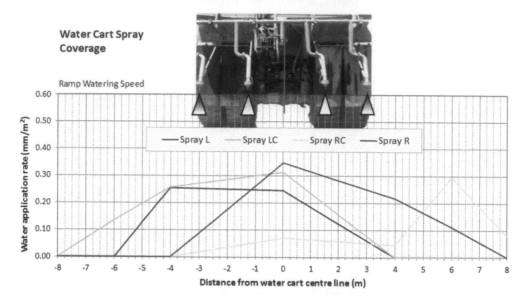

Figure 2.16 Delivery rate determination for water cart sprays.

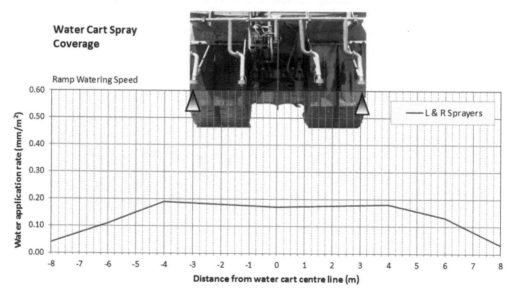

Figure 2.17 Adjusted spray delivery rates to improve coverage and limit over-watering of ramps.

When combining the test methodology described previously with measured applications of water to the road surface, a first indication of the susceptibility of a particular wearing course to the effects of moisture can be determined. Figure 2.19 shows the results from such an assessment, where it can be seen that the wearing course for Test Profile 'A' exhibits an ideal response to wetness or rainfall in that the road can continue to be safely operated over at least 2 mm or more rainfall or water spray, in all likelihood thereby offering true '24x7' trafficability. Test Profile 'B' on the other hand represents a wearing course that is

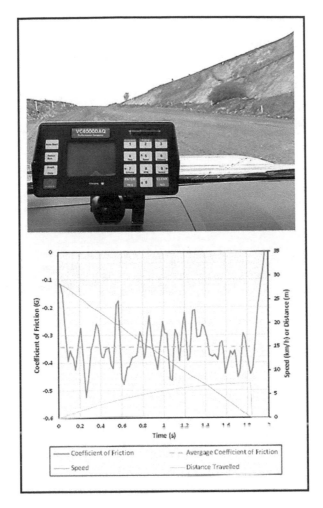

Figure 2.18 Skid resistance measurement using Vericom VC4000DAQ (top) and typical friction testing result (bottom) showing average coefficient of friction measured as 0.34.

susceptible to moisture, showing a significant loss of friction with rainfall (or water-spray) to 0.3 mm/m² and in excess of 2 mm/m² rainfall, the road would be hazardous for mine haul truck operation. In such cases haulage would be suspended until conditions dried back to a point delivering an adequate level of friction commensurate with the geometric design value adopted.

2.3.3 Sight distances

A truck driver needs to be able to perceive a road hazard or obstacle and decide on a course of action. The faster a vehicle is moving, the greater the distance ahead that the driver needs to both see and analyse. This is referred to as the sight distance, as measured from the operators' cab and from the operator's height of view.

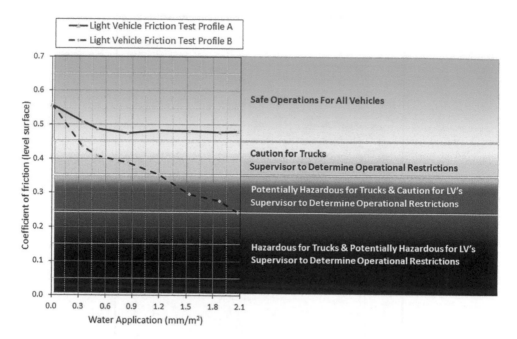

Figure 2.19 Determining the susceptibility of a wearing course to the effects of wetness. Profile 'A' ideal whilst Profile 'B' indicated wearing course skid resistance susceptible to the effects of moisture

Source: operational warrants adapted from the Tulloch and Stocker, 2011 model.

The concept is illustrated in Figure 2.20, in which (initially), approaching a horizontal curve, the bench slope projection limits the driver's sight distance round the curve. Note that, at the point where the truck is located, the driver cannot see the hazard. At this point also, if the truck is to stop before encountering the hazard, the driver should apply the brakes. Obviously, this will not happen since the driver's sight distance is restricted and by the time the obstacle is perceived, sight distance is much less than stopping distance, so the vehicle cannot stop in time to avoid the hazard. Figure 2.21 shows the equivalent situation for a vertical (crest) curve. In both cases, the appropriate geometric alignment of the curves will ensure that sight distance is always maintained in excess of the required 'worst-case' stopping distances. In any instance where sight distance is reduced below the worst-case stopping distance, sight distances should be increased or alternatively speed limits commensurate with the available sight distance should be applied.

Cognisance should be taken of the traffic rules at the site (driving on the left or right) and the position of the driver, which is commonly on the left side of the vehicle for most large mining trucks. Smaller ADTs have a more centralised driving position. Many mines, even where the public drive on the right, adopt a driving on the left rule since drivers can better position their truck when they are looking directly down onto the roadside berm as opposed to only estimating its location off the RHS of the truck.

To address the horizontal curve sight distance problem (Fig. 2.20) by geometric redesign, a 'layback' can be applied, or the horizontal curve radius increased, to increase sight distance to at least the required stopping distance. When the road curves round a bench edge,

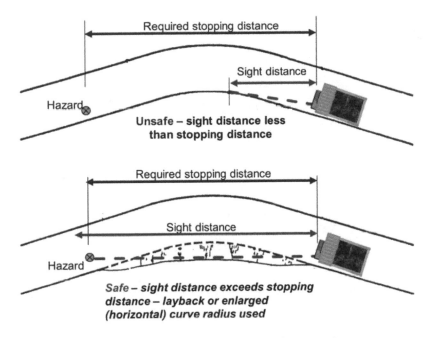

Figure 2.20 Sight versus stopping distance requirements – horizontal curve.

to maintain sight distance a 'layback' (*LB* [m]) is used to keep the road away from the sight obstruction. The layback is found from consideration of the truck minimum stopping distance (*SD* [m] and curve radius of inside lane *R* [m]) following equation 2.5:

$$LB = R\left[1 - \cos\left(\frac{28.65SD}{R}\right)\right]$$

Equation 2.5

Length (*L* [m]) of vertical (crest) curves can be determined from consideration of the height of the operator's eye level, in the cab (h_1 [m]), an object of height (h_2 [m]) (usually a maximum of 0.15 m to represent a prostrate person in the road, spillage or other hazard), *SD* the minimum stopping distance (m) and ΔG the algebraic difference in grades (%) following Equation 2.6 and Equation 2.7.

Where stopping distance is greater than the length of a vertical curve (*SD* > *L*), then:

$$L = 2SD - \left(\frac{200\left(\sqrt{h_1} + \sqrt{h_2}\right)^2}{\Delta G}\right)$$

Equation 2.6

Where stopping distance is less than the length of the curve (*SD* < *L*):

$$L = \left(\frac{\Delta G\ SD^2}{200\left(\sqrt{h_1} + \sqrt{h_2}\right)^2}\right)$$

Equation 2.7

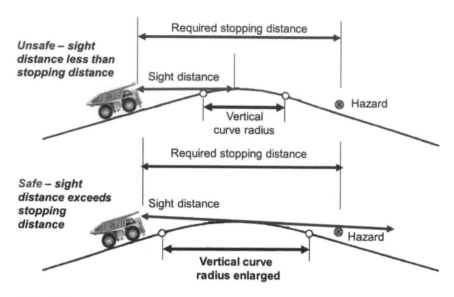

Figure 2.21 Sight versus stopping distance requirements – vertical curve.

Similarly, for the case of the sag curve (where the change in grade is positive, such as dump to flat-haul descents), trucks traveling through sag vertical curves are accelerated upward and the combination of inertia and gravity result in an increased wheel load through the curve and typical vehicle suspension 'bounce' which continues some distance beyond the curve. For both crest and sag curves, a parabola is commonly used since the truck undergoes a constant vertical acceleration, with the length of the sag or crest curve being directly proportional to the grade change. Again, standard survey texts outline the geometric relationships in more detail.

During the daylight hours, sag vertical curves normally present operators with a commanding view of the haul road. However, at night they limit the forward spread of the truck headlights thus the headlight beams (specifically their upward divergence angle (i_d°) and height above ground level (h_l) (m) are considered in sight distance calculations. As before, two options are evaluated with Equation 2.8 and Equation 2.9.

Where stopping distance is greater than the length of a sag curve ($SD > L$), then:

$$L = 2SD - \left(\frac{200\left[h_l + SD\tan\left(i_d\right)\right]}{\Delta G} \right)$$

Equation 2.8

Where stopping distance is less than the length of the curve (SD < L):

$$L = \left(\frac{\Delta G\,SD^2}{200\left[h_l + SD\tan\left(i_d\right)\right]} \right)$$

Equation 2.9

On public roads, passing or overtaking sight distance is also considered to allow vehicles to overtake slower moving vehicles. In the mining environment, overtaking should not be permitted unless a vehicle is stationary or slow-moving, and not without a clear view of the road ahead and after radio call-up and a positive response from the vehicle operator to allow overtaking. Thus, passing sight distance does not need to be considered unless an unusual situation exists.

2.3.4 Truck cab blind spots

In addition to sight distance requirements and the effect of road geometry on these limits, it must also be remembered that in a large mining truck the operator does not have full 360° vision around the vehicle. This is referred to as blind spots and will vary from machine to machine. When evaluating sight distance and, critically, intersection sight distances, it is important to consider whether or not the combination of truck positioning on the road, and the road geometry itself, will facilitate the required sight distance.

A typical driver-seat view from the cab of a large haul truck is shown in Figure 2.22, highlighting the limitations to vision in the vicinity of the truck and especially the RHS off-side position where the ROPS supports and other equipment on the truck terrace (mirror and camera washer bottles and engine fire suppression system) limit line of sight.

This blind spot is shown as part of a typical operators field of view diagram for a large haul truck in Figure 2.23, adapted from the blind area diagrams study (for selected mining vehicles) by CDC NIOSH (2006). The measurement approximates to ISO 5006, Earth-moving machinery – Operator's field of view and indicates both the direction (0–360°) and depth of obscured field of view (in some locations, well in excess of 50 m at ground level).

The location of operator blind spots is important in determining the alignment of the haul road and anticipating where, when negotiating the network of roads, operator blind spots may exist, as illustrated in Figure 2.24. Priority rules should also be considered at the same time as alignment is determined, since this will also affect how road users interact with other vehicles and the alignment. Priority rules are based on which class or type of vehicle give

Figure 2.22 Blind spots from operator cab looking towards RHS of terrace.

way to each other, generally in the form of a hierarchy where typically the following is at play (when driving on the left):

- Emergency vehicles (with siren activated) have right of way over all other vehicles.
- Road maintenance equipment (engaged in grading or watering) have right of way over haul trucks, other heavy vehicles and mobile plant.
- Haul trucks give way to other haul trucks on their left.
- Haul trucks have right of way over other mobile plant.
- Other mobile plant has right of way over light vehicles.
- Light vehicles give way to other light vehicles on their right.

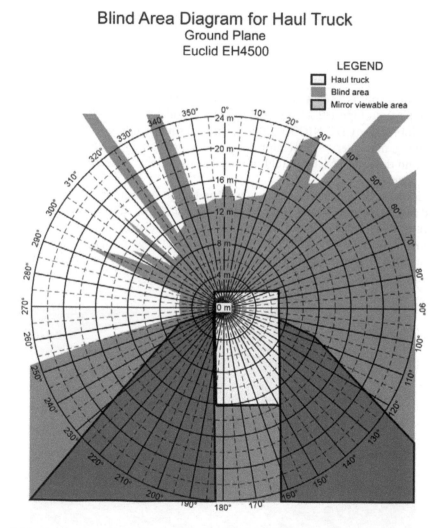

Figure 2.23 Ground-level blind spots for typical large RDT

Source: reprinted courtesy of Centers for Disease Control and Prevention, National Institute for Occupational Safety and Health, USA. https://www.cdc.gov/niosh/topics/highwayworkzones

Figure 2.24 Examples of blind spot interaction with geometric (horizontal) design – limited sight over RHS of truck of any approaching traffic.

Source: Map data: Google, CNES/Airbus 2018.

2.4 Vertical alignment issues – gradeability and brake performance

Vertical alignment is primarily a function of pit geometry and the associated ramp gradients and transitions from grade sections to flat haul. Maximum gradients should be based on gradeability considerations for the type (electric or mechanical drive) and model of truck operated. Each truck has a specific speed-rimpull-gradeability characteristic which is most often related to either total or effective resistance (grade ± rolling resistance) or simply grade (in which case, rolling resistance is added or subtracted as appropriate). Additionally, the brake performance characteristics also need review – although on the return downgrade journey, braking capacity of unladen trucks would theoretically allow high speed haulage, practically it is often necessary and desirable to limit unladen descent speeds according to road conditions, rather than limit purely by brake capacity. With downgrade laden hauling, braking capacity becomes more critical and under certain circumstances, a truck model can be specified with a larger installed braking capacity to enable higher average down-grade hauling speeds to be maintained than would be the case for a 'standard' truck model.

Gradeability data will also indicate the maximum speed of a truck under laden or unladen conditions and where braking is not the limiting speed factor, about 85% of this top speed should be used for design purposes – why slow-up a truck when the engine power was purchased to actually complete a haul in a shorter time? Speed limits will always be necessary under certain operating circumstances on any haul road network, as will be discussed in the following sections concerning geometric design.

Typical standard brake and retarding configurations for a 200 t payload class mechanical drive haul truck with 4.8 kW (SAE gross) engine power/t GVM are shown in Figure 2.25. In the design of the ramp grade for a truck running (uphill) against the grade, limiting drive to D1 would imply a total resistance of 16%, equating to a grade of typically 13% if a rolling resistance value of 3% were assumed. However, a slight increase in rolling resistance or grade above stated values would cause the truck to operate the next lower speed range (torque converter drive) which will accelerate damage to engine and drive-train components as a result. Similarly, to ascend the ramp in D2, a maximum of 12% total resistance would limit the actual ramp grade to 9% if 3% rolling resistance were assumed. The same methodology can also be applied for an unladen truck working against the grade, but in this case selecting the intersection of Total Resistance with Empty Vehicle Mass (EVM) lines.

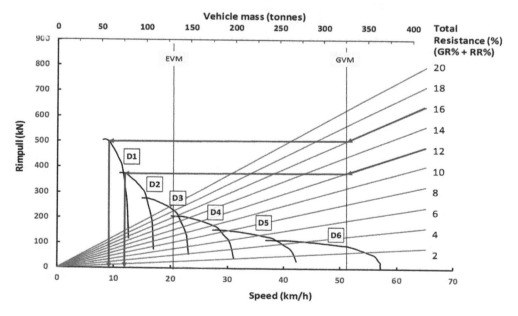

Figure 2.25 Gradeability limits for typical 200 t payload mechanical drive RDT.

In the case of a similar sized and powered electric drive truck there is a theoretically higher limit to gradeability, but as with mechanical drive trucks, with the truck working closer to its physical limits, mechanical and wheel-motor damage would be more severe, in addition to the traction and braking limitations imposed by severely steep grades.

The brake performance of a truck is a key road design consideration especially when the truck is used in a laden-favourable (down-hill) grade configuration. For more conventional laden-unfavourable (up-hill) configurations, brake performance is only considered once the optimal grade has been specified and the impact of this decision analysed on unladen truck speed and road geometry. In this case, the effective total resistance is the ramp grade minus the rolling resistance.

A typical brake performance chart for a truck is shown in Figure 2.26 for a similar mechanical drive truck as was illustrated in Figure 2.25. The truck will descend the ramp in a gear that maintains engine rpm at the highest allowable level, without over-revving the engine. If brake cooling oil overheats, speed is reduced by selecting next lower speed range. With electric-drive trucks the braking effect is achieved through retard and mechanical braking. When using this information for design purposes, it is important to select the appropriate braking grade distance chart that covers the total downhill haul, not individual segments of the haul. Referring to Figure 2.26, for the ramp at 13% grade, subtracting the rolling resistance of 3% gives an effective grade of 10%, in which case the truck will descend (laden) in D2. In the case of the 9% grade, the truck would descend in D3. This data is based on 32°C ambient temperature at sea level with recommended tyres and OEM recommended GVM. This requirement will vary with the truck model and brake options selected. If the actual maximum safe speed of the truck under retard or braking is not exceeded, then speed limits

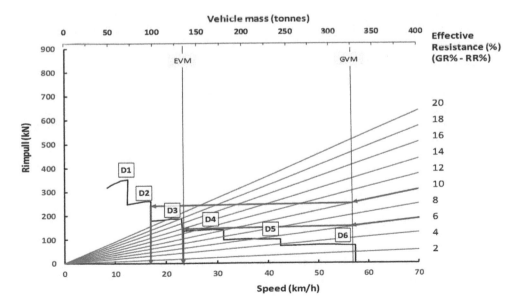

Figure 2.26 Continuous grade braking limits for typical 200 t payload mechanical drive RDT.

may be necessary under certain circumstances, as will be discussed in the following sections concerning geometric design.

Whilst maximum gradients may be limited both by local regulations and vehicle performance considerations, ideally the gradient should be a smooth, even grade over the length of the ramp, not a combination of various grades (containing several grade 'breaks'). Laden trucks running against the grade work best at a total (effective) (i.e. grade + rolling resistance) value of about 8–11%. However, each truck engine and drive system combination has a characteristic 'optimal grade curve' and it is a good geometric design starting point to determine the optimal gradient for the selected truck in use at the mine. It should be noted that whilst travel times (laden) are sensitive to grades against the load, care should also be taken when selecting the grade, from the perspective of truck retard (braking) limitations on the unladen downward leg of the haul. This aspect becomes critical in the case of downgrade laden hauling when on-board retard capacity would be the limiting design criteria.

The optimal grade for a particular truck, engine and drive system option lies between these two extremes:

- A long shallow ramp (low grade) – (truck is fast because effective grade is low, but ramp is long – hence long travel times)
- A short steep ramp (high grade) – (truck is slow because effective grade is high – but ramp is shorter – hence long travel times)

Note that, as mentioned earlier, the optimal grade of the ramp selected using the approach below does not necessarily deliver the most appropriate grade from a mine-planning and resource-optimisation perspective.

In the example shown in Figure 2.27, a simulation was used to determine optimum grade curve for a particular truck with between 2% and 5% rolling resistance (RR) added to the grade resistance. The truck travel time is minimum (or rate of ascent at maximum) at approximately 11% grade (at 2% RR), about 310 s for a 100 m vertical rise. But at the higher grades, the truck 'works' harder and will be more expensive to operate and life-cycle costs may be adversely affected. Also take note of the assumptions used in the simulation work – especially length of ramp, curves or switchbacks and speed of truck on entry to the ramp. As RR increases, the optimum grade will decrease by the same amount.

At grades other than the optimal grade, it is also worth investigating the change of speed associated with changes in grade and rolling resistance. Depending on the specific truck type and drive system adopted, it is not always a smooth change in the rate of ascent with grade (or increasing rolling resistance at a certain fixed grade). This effect is seen in Figure 2.27 where for instance the rate of ascent at 7% grade (with 2% rolling resistance) is less than at 6% grade with the same rolling resistance.

On the (unladen) down-grade haul, small increases in rolling resistance do not play a significant role in the down-grade unladen travel times since these generally favour retard-capacity limited speeds.

2.5 Horizontal (longitudinal) alignment issues

2.5.1 Width of road

Pavement (road) width should be sufficient for the required number of lanes. The associated safety shoulders are incorporated in the carriageway width and drainage and other associated

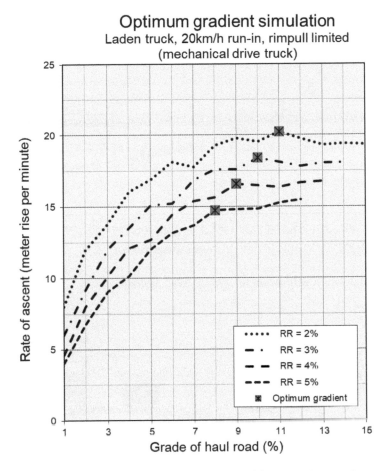

Figure 2.27 Optimal gradient determination for a typical large mining truck.

features should be included in the formation width. The widest vehicles regularly in use at a site will determine the road width. Width of typical RDTs can be estimated for preliminary road width and associated formation or reserve requirements from the Equation 2.10:

$$w = payload^{0.38}$$

Equation 2.10

As shown in Figure 2.28 where
w = Truck width (across body) (m)
payload = Payload (t)

At least 3.5 times the width of the truck (w) should be used for the road width for bi-directional travel and in cases where the truck fleet is expanded with larger vehicles, this can be problematic when existing roads, designed for smaller payload vehicles, are considered.

Figure 2.29 shows a lane width of 13 m and a road width of 23 m for a 6.5 m wide, 137 t payload RDT. This width excludes shoulders, berms (roadside and median) and drains, etc.

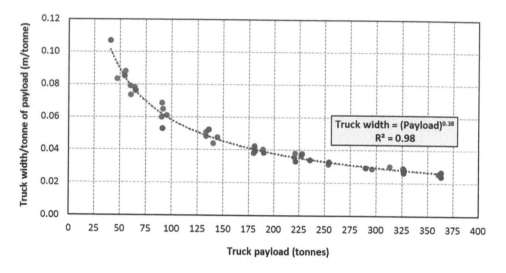

Figure 2.28 Truck width estimation for typical RDT payloads.

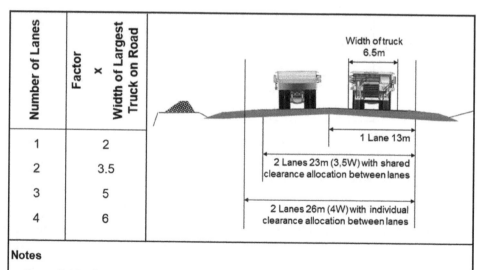

Figure 2.29 Determination of road width based on widest vehicle regularly using the road.

Note that this accepted design methodology (3.5w) requires 'sharing' of the clearance allocation between lanes, which will require good driving skills – especially with larger haul trucks (to judge off-side clearance if driving on the LHS of the road or the location of the roadside berm if driving on the RHS). Where traffic volumes are high or visibility limited, a safe road width would be 4 w.

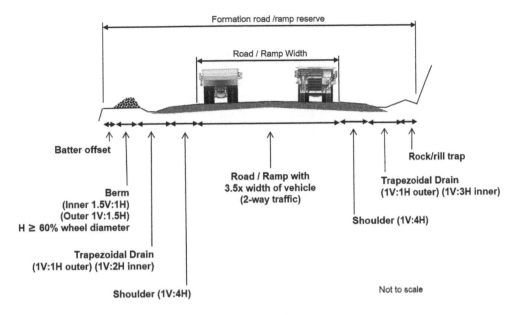

Figure 2.30 Typical road formation for a 260 t payload RDT.

A typical design for a high traffic volume haul road running 260 t payload haul trucks is shown in Figure 2.30 (with optional median berm, as discussed later). Minimum road/ramp reserve in this example is 48 m, from outslope bench crest to inslope bench toe. This width comprises the following:

• Outslope batter offset (crest erosion, blast damage, etc.)	2.3 m
• Outslope berm (refer berm design notes)	4.7 m
• Outslope trapezoidal drain (refer drainage notes)	4.0 m
• Outslope shoulder	2.0 m
• Road width (3.5w @ w = 8.3m)	29.0 m
• Inslope shoulder	2.0 m
• Inslope trapezoidal drain (refer drainage notes)	4.0 m

Since road width can critically impact both stripping ratio and ore recovery, reducing the formation or road/ramp reserve width is often seen as a means to improve the economics of a mining operation. However, any reductions considered must be evaluated against the safety implications of such reductions. Batter offset can be reduced/omitted where crest damage and/or water run-off erosion is minimised and blasting practices reduce potential crest insta-bility/damage. The rock trap can be omitted where good quality blasting is used and limited bench rock-fall is anticipated. Shoulders can be omitted when traffic volumes are low, but this may cause trucks to run with their wheels in or close to roadside drains and, if vehicles park or become unserviceable on road, temporary road traffic control will be required for safe traffic operations whilst removing the obstacle. Safety statistics have also shown that, especially with high traffic volume roads, omitting shoulders is associated with an increased risk of collision with stationary vehicles. No inslope berm is used unless embankment batter slopes exceed 1V:4H, road side drop-offs are greater than 0.5 m or there is an unacceptable risk of collision between vehicles or road side obstacles.

2.5.2 Turning circle of large haul trucks

The starting point for any horizontal curve design would be from consideration of the turning circle of the largest, or limiting, vehicle intended to use the road. The turning path of a typical two-axle rear-dump haul truck is based on the outer front (steering) wheel path following a circular arc at under 16 km/h. GVM, tyre and suspension characteristics have a negligible effect in low-speed turns. However, consideration should be given to the swept path or required additional clearance due to the combination of turning circle and innermost and outermost projections of the vehicle body when negotiating a curve and any off-tracking, as shown in Figure 2.31.

The following definitions apply:

- Turning circle clearance radius (machine clearance diameter ISO 7457:1997), (turning radius – wall-to-wall SAE J695:2011) – radius of the smallest circle which will enclose the outermost points of projection of the machine and its equipment and attachments when it executes its sharpest practical turn.
- Turning circle radius (ISO 7457:1997) – radius of the circular path described by the centre of tyre contact when the wheel describes the largest circle when the machine is executing its sharpest practicable turn.
- Turning centre – point about which all turns of constant radius are made. For ideal steering, free of tyre scrubbing, the extended axis of all wheel spindles passes through this centre. In the case of dual assemblies in which the axles are constrained to parallelism,

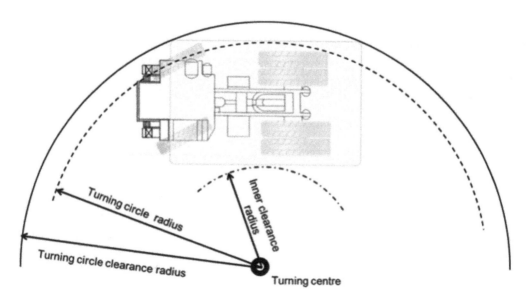

Figure 2.31 Turning circle for RDTs.

the turning centre is assumed to fall on a line parallel to and midway between these axle centrelines.

- Turning diameter – double the turning circle radius described previously.

Off-tracking refers to the difference in the centre-point paths of consecutive axle paths which are not coincident and where an offset exists between axle paths. The amount of off-tracking (D_{ot}) (when negative implies tracking inward towards turning centre) that will affect the swept path (difference between inner and outer clearance radius) can be determined from this equation:

$$D_{ot} = -R + \sqrt{R^2 - WB^2}$$ Equation 2.11

Where
R = Radius of curve (m)
WB = Wheel base of truck (m)

For example, off-tracking for a 220 t class of RDT is approximately at a maximum of 1.1 m for the minimum turning circle clearance radius of 16.5 m, as shown in Figure 2.32. Practically, it is often necessary to select a lowboy or flat-bed truck-trailer combination as the limiting vehicle.

2.5.3 Curvature and switchbacks

Any curves or switchbacks should be designed with the maximum radius possible (generally >200 m ideally) and be kept smooth and consistent. Changes in curve radii (compound curves) should be avoided. A larger curve radius allows a higher safe road speed and increased truck stability. Sharp curves or switchbacks will increase truck cycle times, haul costs, tyre costs and require additional road maintenance. Ideally, when joining two straights with a

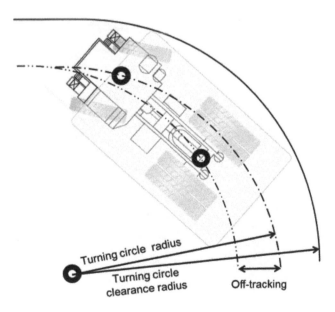

Figure 2.32 Off-tracking for a 220 t payload RDT.

circular curve, a clothoid transition curve should be used. A clothoid curve has a constant rate of change of curvature, which gives a constant rate of change in lateral acceleration at constant speed. Basic clothoid geometric relationships can be found in many standard texts on surveying.

The dual tyres on drive axles are especially prone to wear going around tight curves. A switchback with an inside depression dug from tyre slip is common and if the depression exposes road base, these rocks will damage the tyre, as shown in Figure 2.33. However, many truck models now offer a 'dual-differential', which allows for different dual tyre rotation speeds, and which reduces the impact of tight curves on tyres (but importantly, not on the road and, critically, the wearing course). These enhancements improve the service life of the differential and dual wheel components where tight radius curves and switchbacks are numerous.

Minimum curve radius (R [m]) can be initially determined from:

$$R = \frac{v_o^2}{127(U_{min} + e)}$$

Equation 2.12

Where

e = Super-elevation applied (m/m width of road)
U_{min} = Coefficient of lateral friction supply
v_o = Vehicle speed (km/h)

U_{min}, the coefficient of lateral friction supply, is generally taken as 0.0 (where no measured data available) to a maximum of 0.1. Where the pit layout requires a tighter radius than the minimum radius indicated at a particular truck speed, speed limits need to be applied on approach to the curve. Care should also be taken when assuming a value for the coefficient of

Figure 2.33 Road base exposed on switchback, with potential to damage truck tyres.

lateral friction – many of the factors that influence the coefficient of longitudinal deceleration (as discussed in Section 2.2.4) also apply to this coefficient.

Switchbacks should be designed with an inside ramp string radius to give a minimum inside tyre path radius of at least 150% of the minimum turning circle clearance radius. Width of switchbacks including inslope and outslope berms and inslope small trapezoidal drain will vary with road width design as described previously. Outslope berms can be omitted if safe to do so but the inslope berm should be built to assist the truck drivers sighting alignment through the switchback. Divider or median berms should be used where the possibility of truck sliding exists, but subject to the limitation imposed by a low-bed or float if used at site.

Switchbacks should be designed ideally at no more than 3% gradient for at least the semi-circumferential length of the switchback with super-elevation applied throughout the curve. However, high rates of road wear and degradation can be anticipated on and leading down to these features due to the tight radius curve, scouring and vehicle braking requirements on approach. If designed on grade (which is less ideal, but often an operational requirement), as shown in Figure 2.34, a maximum 10% gradient on the inside curve radius should be maintained and speed limits should be applied well in advance of the switchback for downgrade vehicle travel (and invariably, significant wearing course damage will be seen on the down-grade approach to the switchback as a result). Drainage should be led off the road on the upslope side of the switchback since drainage in the neck of the switchback is limited and problematic. Drainage on the outside of the curve is not required when super-elevations are applied. A typical design is shown in Figure 2.35 (based on a given truck width and associated drainage and safety features).

2.5.4 Curve super-elevation (banking)

Super-elevation refers to the amount of banking applied from the outside to the inside of a curve to allow the truck to run through the curve at speed (generally when speed exceeds

Figure 2.34 Truck on approach to switchback.

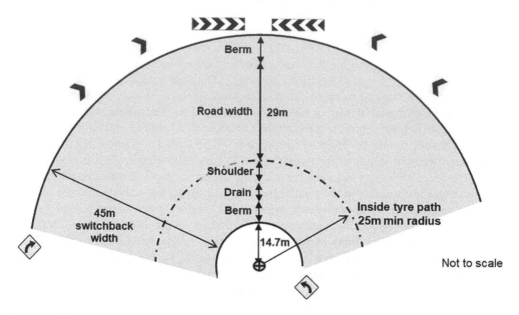

Figure 2.35 Schematic layout of switchback and associated signage.

about 15 km/h). Ideally, the outward centrifugal force experienced by the truck should be balanced by the lateral friction supply between tyres and road. Super-elevations should not exceed 5–7%, unless high-speed haulage is maintained and the possibility of sliding minimised. In Figure 2.36 with no super-elevation applied in the curve, the truck centrifugal and centripetal forces are unbalanced and tyre positions 1, 3 and 4 (as defined in Figure 3.16) (i.e. LH front [LF], LH Rear-outer [LRO] and -inner [LRI]) carry higher loads, leading to potential tyre and strut/chassis stress issues (as well as damage to the road surface, spillage and driver discomfort).

Under the ideal combination of speed, radius and super-elevation, there should be nearly zero excess side or lateral frictional forces. When the conditions of speed, radius and super-elevation are not matched, side frictional forces are generated, leading to accelerated tyre abrasion, and, should available side friction be exceeded, truck sliding. Tyre manufacturers generally recommend side friction does not exceed approximately 0.05 to 0.1.

Table 2.1 shows typical super-elevation rates based on speed of vehicle and radius of curve (the radius of the centreline of the road lane on the inside of the curve – the radii of the inside shoulder, centreline and outside shoulder are found from consideration of largest operating haul truck and road width values) with U_{min} (coefficient of lateral friction supply) set to zero. Alternatively, Equation 2.12 may be used to determine the super-elevation (m/m width of road) where U_{min} varies up to a maximum of 0.1.

Elevation rates in excess of 5% should only be applied as a combined super-elevation with a road (median) splitter berm used to separate slow and fast lanes of the road (each with its own speed-related super-elevation), due to the possible instability of slow-moving vehicles negotiating higher rates of super-elevation (especially where the road is wet). Where tighter curves are required or truck speed is higher on approach to the curve, a speed limit should be applied.

Figure 2.36 Truck path through a curve designed without super-elevation.

Table 2.1 Recommended rates of super-elevation for mine roads.

Curve Radius (m)	Speed (km/h) and super-elevation (m/m width of road)								
	15	20	25	30	35	40	45	50	55
50	0.035	0.060	0.090						
75	0.025	0.045	0.070	0.090					
100	0.020	0.035	0.050	0.075	0.090				
150	0.020	0.025	0.035	0.050	0.065	0.085			
200	0.020	0.020	0.025	0.035	0.050	0.065	0.080		
300	0.020	0.020	0.020	0.025	0.035	0.045	0.055	0.065	0.080
400	0.020	0.020	0.020	0.020	0.025	0.035	0.040	0.050	0.060
500	0.020	0.020	0.020	0.020	0.020	0.025	0.030	0.040	0.050

Notes
Super-elevation rates of 0.01–0.05 operable under most conditions.
Super-elevation rates of 0.05–0.09 operable under limited circumstances and should be accompanied by a median berm to separate traffic on both sides of road.
Superelevation rates in excess of 0.1 are not recommended.

2.5.5 Road camber – cross-fall or crown

The use of a road camber is critical to the design and successful operation of mine roads. Applying a camber ensures water does not gather on and penetrate into the road surface. Standing water on or in an unpaved haul road is extremely damaging and every attempt should be made to get water off the road as quickly as possible – but without inducing excessive erosion caused by high run-off velocities.

Two camber options are generally recognised, as shown in Figure 2.37:

i.   cross-fall, where the camber runs across the full road width, from a high to a low point (and associated drain) as shown in Figure 2.38, or
ii.   crown, where the camber runs from a high point in the centre of the road to low points on either side of the road, each with its own drain.

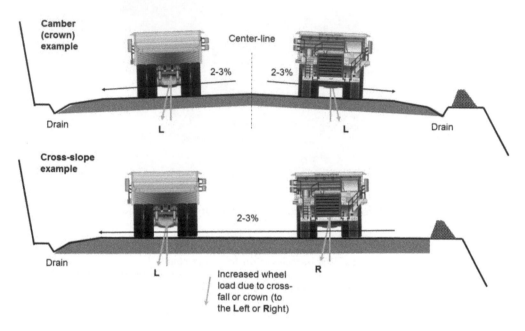

Figure 2.37 Cross-section options.

Figure 2.38 Example of cross-slope applied to a ramp road, with drainage at the inslope (toe) of the bench.

Whatever option is adopted, at the point where the road edge and camber down-slopes meet, a table drain or drainage ditch must be provided. It is critical to ensure that the drain forms part of the road formation, and is well compacted, to prevent run-off water from simply penetrating through the drain and into the layerworks. Furthermore, as will be discussed in Chapter 4, the camber is established in the layerworks to ensure that layer thicknesses are maintained throughout the structure and not simply cut into the wearing course alone.

On ramp roads, a reduced amount of camber can be used. If absent, however, water will only run down the ramp and not down and off to side drains. If water doesn't run off the road, it will eventually gather enough speed to cause serious erosion of the wearing course and drains at the low-point of the ramp.

A cross-sectional slope (crown or cross-slope) of 2–3% is ideal, providing adequate drainage without incurring adverse truck tyre and strut loading conditions. A preference may exist for cross-slopes due to the envisaged equalised load sharing and reduced tyre scrub (as shown in Fig. 2.37). Tyre temperature monitoring confirms that the crown option leads to the wheel positions 1, 3 and 4 running at slightly higher temperatures as a result of the unequal load sharing. The cross-slope option is best used where the slope falls to the inside of the bench (toe or inslope position). If used when the slope falls towards the outside of the bench (crest or outslope position), extreme caution is required as this leads to the possibility of trucks sliding in the direction of the bench crest or outslope, or towards a large vertical drop – large deflection berms/windrows should be placed at the road edge under these circumstances.

Special consideration must be given to determining when to use the maximum and minimum rates of cross-slope or crown. Lower rates are applicable to relatively smooth, compact road surfaces that can rapidly dissipate surface water without the water penetrating into the road surface. In situations where the surface is relatively rough, a larger cross slope is advisable. On well-constructed gravel and crushed rock roads, with a longitudinal grade of more than 3%, the 2% criterion is preferable. Excessive slopes lead to erosion of the wearing course – which tends to be more prominent at the outer edges of the road (due to the higher run-off velocity) – and often coincident with the outer tyre path of the truck. Care should be taken with higher rates of camber in conjunction with steep longitudinal grades, the combination of which can cause a vehicle to slide.

2.5.6 Development of super-elevation

A run-in or run-out refers to a section of haul road used to change from a normally cambered (cross-fall or crown) into a super-elevated curve section and back again. The change should be introduced gradually to prevent excessive twisting or racking of the truck chassis. The run-in or -out length is typically apportioned 25–34% to the curve and 66–75% to the tangent or run-in or -out to the curve. Typical examples are shown in Figure 2.39 for the case of crown and cross-fall sections.

Run-in or -out lengths vary with vehicle speed and total cross-fall or crown change and can be estimated from the equation below where CSx is the maximum percentage change in cross-fall per 10 m-road length and v_0 the speed of the truck (km/h), as in Equation 2.13 or Table 2.2:

$$CSx = 3.33 - 0.0274 v_0 \qquad \text{Equation 2.13}$$

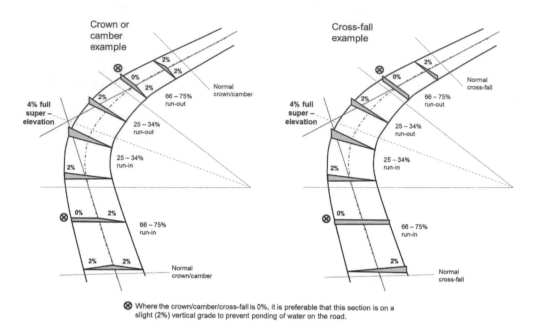

⊗ Where the crown/camber/cross-fall is 0%, it is preferable that this section is on a slight (2%) vertical grade to prevent ponding of water on the road.

Figure 2.39 Transition options for crown and crossfall run-in and -out.

Table 2.2 Run-in and run-out application rates at a range of typical truck speeds.

Vehicle speed (km/h)	10	20	30	40
CSx (in m/m width of road) over 10 m length of road	0.030	0.028	0.025	0.022

Generally, 0.025 m/m/10 m length of road is a good rule of thumb for the maximum run-out rate that should be used. Where the run-in or run-out is at 0% (i.e. the road is 'flat') there should be a slight longitudinal grade to prevent water from ponding on the road at this point. These transitions can also be incorporated in a mine road design by 'eye' or 'feel' rather than by calculation – if the curve and approaches feel safe and comfortable in a light vehicle, it will also be suitable for a large mine haul truck.

2.5.7 Combined alignment

A few tips when laying out a road with all the factors discussed previously – to prevent some of the more common geometric design problems often encountered.

- Avoid sharp horizontal curves at or near the top of a grade section of road. If a horizontal curve is necessary, start it well in advance of the vertical curve.
- Avoid switchbacks where possible – but if the mine plan dictates their use, make radius as large as possible, open road to 4x width of largest truck and avoid placing on grades >3%.

- Avoid sharp horizontal curves requiring a (further) speed reduction following a long-sustained downgrade where haul trucks are normally at their highest speed. Harsh braking before the curve will always generate excessive wearing course damage.
- Avoid short tangents and varying grades, especially on multi-lane roads. Grades should be smooth and of consistent grade percentages.
- Avoid intersections with poor drainage. Drainage design at intersections should prevent any ponding of water against intersection.
- Avoid sections of road with no camber or cross-fall. Often encountered at curve super-elevation run-in or -out, these flat sections should preferably be at a 1–2% vertical grade to assist drainage.
- Avoid staggered crossroads or other multiple road junctions. Preference should be given to three-way over four-way intersections. Re-align roads to provide for conventional cross road layouts and at any junction, always provide splitter or median islands to prevent vehicles cutting corners through a junction.
- Avoid vegetation, roadside furnishings or excessively high splitter islands that would otherwise eventually limit sight distances in any of the four quadrants required, especially in the case of other road users. Take care to locate signage so as not to obstruct sight lines.
- Avoid having the inside (and lower) side of a super-elevated bench-to-ramp access road at a steeper gradient than the ramp road itself, by reducing the centreline grade of the curve. The inside grade of the curve should not exceed that of the ramp road. Using a transition spiral, and where pit room permits, set the inside gradient of the curve flatter than the ramp grade by 2–3% to compensate for increased curve rolling resistance.

2.6 Safety berms

The function of a 'crest' or road-edge berm, bund or windrow can either be considered as an 'arresting' device, or as a 're-directing' device for misaligned trucks. A berm will not effectively stop trucks (especially high-speed laden or unladen trucks) from leaving the road. At best, they will provide limited deflection and warning to the driver that the truck path needs correcting. The material comprising the berm and its natural angle of repose significantly influence how the berm performs. Figure 2.40 shows a typical berm of adequate height, but which has too shallow an inner (road) side face.

The slope of the inner (road) side of the safety berm should be preferably as steep as possible – 1.5V:1H – if needs dictate, by using an engineered or stabilised material since the angle of repose of loose unconsolidated rock will be significantly lower than the 1.5V:1H (56°) recommended here. A steep (inner) berm face ensures better re-direction of the truck and less tendency to climb and topple. But in doing this, ensure stability and maintenance of height because a flat or low berm will also induce truck roll-over. Also note that with this 're-directing' effect, the truck itself may become a hazard on the road when it is redirected by the berm towards oncoming traffic.

Historically, berm heights of 50% of the truck wheel diameter were standard, but with the advent of articulated 4x4 and 6x6 wheel drive trucks and the more recent development of ultra-class rear-dump trucks, the berm height should be at least 60% of the truck wheel diameter. A typical design for a 180 t payload haul truck with 36.00R51 tyres is shown in Figure 2.41 – primarily for re-directing as opposed to arresting. With the latter (arresting) design, a larger outslope batter offset may be required if the truck is anticipated to straddle

Figure 2.40 Typical roadside safety berm – but in need of steeper inner face.

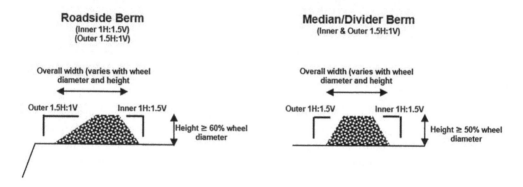

Figure 2.41 Roadside and median berm design for large mine RDT.

the berm during deceleration. In all cases, a >2 m high berm presents various construction challenges, not the least of which is the equipment capable of dumping and shaping the berm at this height, in addition to the large footprint of the berm which requires a considerable increase in road reserve. In many countries, working above 2 m height is considered a risk and anchorage points or other safety features are also required to prevent injury to or falls of workers that may be required to access the berm.

In an attempt to reduce the footprint of roadside berms required for large haul trucks, and especially to ensure a steep inner face for the berm, a prefabricated system of interconnected 'berm walls' have been used at several mine sites. The system consists of a series of filled polyethylene shells which secure together in a row along the edge of an open pit road, as illustrated in Figure 2.42. The wall acts to support a windrow of loose material, presenting a vertical face to trucks interacting with the berm, as well as preventing trucks' tyres from running up the bund. In the event of a collision the truck would either deflect off the berm or is slowed as the berm absorbs the truck's momentum.

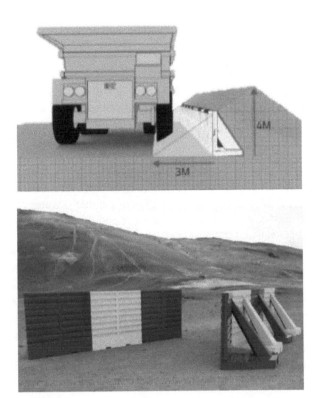

Figure 2.42 Safescape edge protection EP barrier designed to minimise berm footprint at roadside.

Source: Image courtesy of Safescape Pty Ltd.

Truck GVM, speed and approach angle have a significant deformation effect on a conventionally constructed berm, which is typically built from unconsolidated material. The ability of a berm to re-direct reduces as angle of truck approach increases. Furthermore, large tyre sizes and non-centering steering mechanisms reduce the tendency of the truck to redirect itself when encountering a berm. With 4x6 and 6x6 wheel drive articulated dump trucks, berm dimensions of at least 66% of the wheel diameter are recommended, due to the truck's ability to climb smaller berms. Other factors such as inertial characteristics, sprung mass ratio and suspension characteristics indicate significantly different response patterns for haul vehicles when encountering berms.

Where a median (centre) berm is used to split two lanes of traffic, or in the vicinity of junctions (splitter islands), the same design principles should be applied (except for junction splitter islands – reduce the height of the berm so as to not restrict intersection sight and, as is often adopted by mine sites, use discarded and filled truck tyres to delineate splitter islands). Consideration also needs to be given to both the function of the median or centre berm and the implications in using such. In addition to the cost of construction and the additional formation width that is required (which could impact stripping ratios), how to accommodate grader maintenance, broken-down vehicles, low-beds, etc. and the impact on drainage (cross-fall drainage will not be appropriate with a median berm – a crown must be used) should all be additional considerations. Note that this discussion relates primarily to roadside and median berms, the design principles for tip head berms are not addressed here.

2.7 Ditches and drainage

A well-designed drainage system is critical for optimum mine haul road performance. Water on the road or in the road layerworks will quickly lead to poor road conditions. As part of the haul road geometric design process, contours in the vicinity of the proposed road should be examined prior to construction to identify areas of potential ponding, direction of drainage and run-off and the requirements and location of culverts, etc.

The drains at the edge of the road should be designed to lead the water off the road without causing erosion. As discussed previously, a camber is used to lead water off the road to the side drains. Figure 2.43 shows an example of poor camber – water collecting in the middle of the road, not the edges. It is clear that the water ponding in the centre of the road leads to the potholing defect seen and to the subsequent weak spots in the haul road.

Drains should not be cut down into the base layer – ensure drains are 'lined' with compacted material, thereby preventing water from seeping into the underlying layers. Also take care not to leave windrows of wearing course (after grading the road) along the edges of the road – they will also prevent water from draining off the road surface. Make sure that after blading a road, windrows (and if appropriate, safety berms too) are cut through at regular intervals to assist drainage. If circumstances permit, consider blading over-wet wearing course to the centre of the road, not the sides of the road. Windrows of wet material at the side of the road cause ponding of water – and also pick-up spillage which is a problem when opening the wearing course and spreading back onto the haul road.

Figure 2.43 Example of poor camber design with water gathering on the road as opposed to being led off to side drains.

For drainage, V ditches are recommended for nearly all applications, owing to the relative ease of design, construction and maintenance, as illustrated in Figure 2.44. Ideally, drains should be located in undisturbed material rather than fill material. In a cut/fill section, use a cross-slope toward the cut side and run drainage in a single ditch. In a total cut or total fill section, carry drainage on both sides with crown or camber from the road centreline. Side slopes of drains are typically 3H:1V adjacent to road shoulder and should not exceed 2H:1V on the outslope except in extremely restrictive conditions. The outslope will vary with the material encountered. In rock it may approach a vertical slope: in less consolidated material, a 2H:1V slope or flatter.

Drains should be a minimum of 0.3 m deep (unlined) and should be regraded when depth has been reduced by 50%. Where flow capacities require a deeper/wider drain, which may constitute a traffic hazard, the outslope berm should be placed adjacent to the haul road shoulder and be cut-through at regular intervals and/or low points to allow water to drain off the road.

Any drains must be designed to adequately handle expected runoff flows determined by the mine hydrologist/hydrogeologist, under various slope conditions. The primary consideration is the amount of water that will be intercepted by the ditch during a rainstorm. Typically, a 10-year, 24-hour storm chart should govern the design. Ditch lining is a function of road grade and in-situ material characteristics:

- At 0% to 4% grade the drain may be constructed without the benefit of a liner except in extremely erodible material such as sand, dispersive soils or easily weathered shale silts and clays. Spray-seals should be applied to drain and side slopes.
- At grades over 5%, the lining should consist of coarse crushed waste rock (riprap) placed evenly on both sides to a height no less than 0.3 m above the maximum depth, and depending on depth and width of drain, a minimum freeboard of 100 mm should be used where feasible.

Rock lining (riprap) size depends on flow velocity, drain bed slope and rock density. For a typical crushed rock, 2.6 t/m³ may be assumed. Assuming a uniform flow velocity the following sizes (particle size for 50% of the material passing, d_{50}) for angular (crushed) riprap should be used:

- Gradients <5%

    - For flow velocities (v)(m/s) 1.5 > v > 0.5: d_{50} = 100 mm

        - 2.0 > v > 1.5: d_{50} = 200 mm
        - 2.5 > v > 2.0: d_{50} = 300 mm
        - 3.0 > v > 2.5: d_{50} = 400 mm

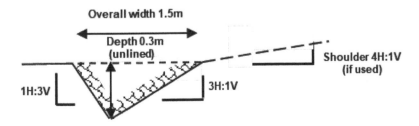

Figure 2.44 Typical V ditch drain.

- Gradients >5%

  - For flow velocities (v)(m/s) 1.5 > v > 0.5: d_{50} = 100 mm

    - 2.0 > v > 1.5: d_{50} = 300 mm
    - 2.5 > v > 2.0: d_{50} = 400 mm
    - 3.0 > v > 2.5: d_{50} = 500 mm

Note that where large riprap is required, depth and width of drain may need to be increased to maintain design capacity.

Minimum thickness of rock lining is recommended as

- $1.4d_{50}$ when d_{50}/d_{90} > 1
- $1.6d_{50}$ when d_{50}/d_{90} > 0.8 (upper limit of quarry rock)
- $1.8d_{50}$ when d_{50}/d_{90} > 0.67
- $2.1d_{50}$ when d_{50}/d_{90} > 0.5 (lower limit of quarry rock)

Velocity of flow should be no less than 0.5 m/s to prevent excess sedimentation. Silt traps or combined energy dissipaters should be considered when flow velocity typically exceeds 4.5 m/s.

Culvert sections, such as that shown in Figure 2.45, are used to conduct run-off water from drainage ditches under the haul road. If buried piping is used, set to 3–4% fall and use smooth-wall concrete pipes in conjunction with a drop-box culvert of a size suitable to enable it to be cleaned with a small backhoe excavator. At all culvert inlets, a protective encasement or 'headwall' consisting of a stable non-erodible material should be provided. Minimum culvert internal diameter is 600 mm and preferably 900 mm for ease of maintenance. However, practical experience often suggests that culverts are liable to silt up quickly where significant amounts of run-off and scour are encountered and maintenance is often impractical.

Typical culvert units are either portal and rectangular precast concrete culvert units or precast concrete pipe culvert units. Depth of cover over the culvert pipe is determined by the type of culvert in relation to the vehicles that will use the road. A minimum cover of typically 500 mm over the pipe is required for the strongest reinforced concrete pipe (Class 10) for most large truck applications, increasing with lower class pipes in most cases. All prefabricated culverts should be constructed under trenched conditions once the road has been constructed. Concrete pipe culverts are laid on a layer of fine granular material, 75 mm thick, after the bottom of the excavation has been shaped to conform to the lower part of the pipe. Where rock, shale or other hard material is encountered on the bottom of excavations, culverts should be placed on an equalising bed of sand or gravel. Once placed, the culvert trench is backfilled and compacted.

A suitable headwall should be constructed from 2 × 1 × 1 m gabions and 2 × 1 × 0.30 m Reno mattress. Gabions are rectangular woven wire mesh baskets filled with rock to create flexible, permeable structures for erosion protection. Reno mattresses are thin, flexible rectangular mesh cages filled with rock to limit movement during high-flow conditions. Because of their flexibility, a Reno mattress is used mainly for scour protection and embankment stability in channel linings.

Under extremely high rainfall intensities, rolling dips (broad-based dips or oblique drains, as shown in Fig. 2.46) should be considered on all roads to intercept water flowing down any ramp road. A rolling dip is generally constructed at the top of the ramp, prior to commencing descent, if there is possibility of water entering the ramp from an intersection

Figure 2.45 Typical 900 mm culvert installation under trenched conditions following haul road construction.

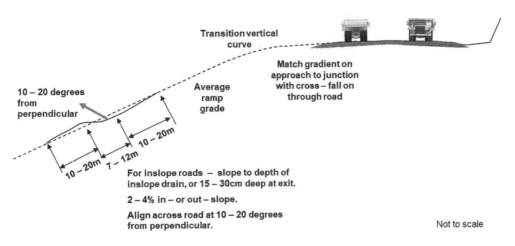

Figure 2.46 Rolling dip installation at top of ramp to intercept runoff from main haul road.

with other (through) haul roads, and further rolling dips should be constructed at 300–600 m intervals on the ramp to catch and disperse surface water (and inslope drain flows if necessary) to the outslope.

In most cases, rolling dips should be angled at 10–20° from perpendicular to the road, to the outslope, with a 2–4% outslope (similar to crossfall applied) and be long enough to allow safe passage of vehicles (at least 50 m). A 3–6% reverse slope should be used. Rolling dip angles more than 20° from perpendicular will enhance water flow velocity and minimise sedimentation but can cause excessive vehicle racking (twist of the truck chassis across diametrically opposite points, e.g. front left and rear right, etc.) and will also increase cycle times due to the requirement to reduce speed on approach. A stockpile of wearing course material should be located on the downslope side of the dip for repair and maintenance of the dip, which may be subject to scour under high intensity rainfall events.

2.8 Intersection design

The predominant type of incident that occurs at intersections of mine roads is due to haul truck and light vehicle interactions in and around the intersection. A key issue is both vehicle misalignment in the intersection and the blind spots of the truck (especially over the terrace to the RHS of the truck). Ideally, the angle at which intersecting roads meet should be between 70° and 90°. In Figure 2.47, a poorly designed and aligned intersection is shown.

Figure 2.47 Poorly designed multiple-road junction.

Source: Map data: Google, DigitalGlobe/AfriGIS (Pty) Ltd 2018).

With the design of intersections, it is important as far as possible to avoid intersections near the crest of vertical curves or sharp horizontal curves (with high super-elevation), or placing an intersection on the inside of a horizontal curve. Intersections should be as flat as possible with sight distances being considered in all four quadrants, and where an intersection lies at the top of a ramp, consider 30–60 m of level road before the intersection and avoid stopping and starting a laden haul truck on grade. The use of splitter islands on the approach to an intersection will help channelise traffic and reduce the tendency to cut corners. It will also assist in demarcating 'no overtaking' areas on the approach to an intersection.

Sight distance requirements at intersections have been summarised by Vagaja (2010) and should be considered from the following perspectives:

- Approach sight distance: the truck operator should have adequate distance to assess, react (and if necessary), stop the truck before entering the intersection areas of potential conflict or for a truck operator on the 'main' haul road to stop the truck should a potential conflict arise with traffic entering from the intersection.
- Minimum gap sight distance: the distance required for truck traffic to join the (main ramp, for example) haul road without impeding traffic approaching the intersection on the main ramp road.

A proposed layout is shown in Figure 2.48, following an original scheme proposed by Vagaja (2010) as applied to LH drive traffic schemes. In the design and especially signage used at the junctions, consideration must be given to traffic rules and the local traffic management plan used at site, especially with 'yield to heavy vehicle' rules, etc.

Roundabouts, as an alternative to more complex multiple entry- and exit-road junctions, should be positioned in safe locations away from horizontal and vertical alignment changes.

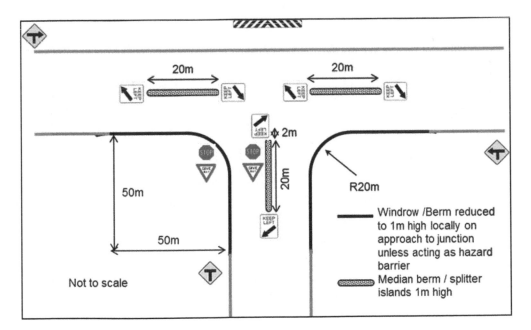

Figure 2.48 Typical 'T' intersection design and signage for haul roads (driving on LHS).

Traffic splitter islands should be installed to slow speeds of vehicles entering the roundabout with Keep Left signs installed at the end of islands. Haul truck tyres painted white may be used as an alternative – but note should be taken of the increased road hazard risk associated with this option.

The design speed of the roundabout is determined from the smallest radius along the fastest allowable path. The smallest radius usually occurs on the circulatory roadway as the vehicle curves to the right around the central island. However, it is important when designing the roundabout geometry that the radius of the entry path (i.e. as the vehicle curves to the right through entry geometry) not be larger than the circulatory path radius and ideally, to maintain sight distance and avoid blind spots, approach the circulatory road at close to right angles. Design software such as ARNDT (Queensland Government Department of Transport and Main Roads, DTMR, 2006) can be used to develop a design suitable for the road junction geometry and vehicles using the roads, a typical example of which (with associated safety signage) as applied to LH drive traffic schemes is shown in Figure 2.49.

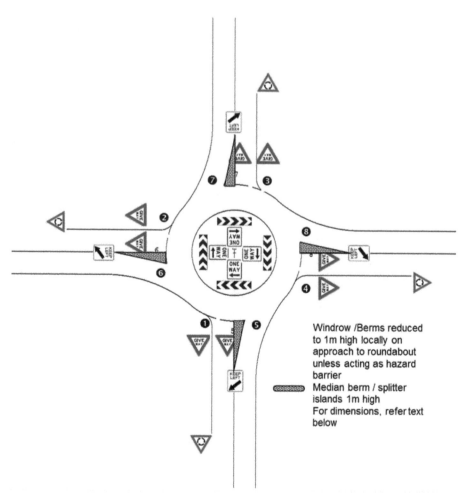

Figure 2.49 Typical roundabout design and signage for haul roads (driving on LHS).

Finally, the demarcation of the road should be considered integral to the geometric design process. Guide posts should be installed to demarcate the edge of road formations and indicate the alignment of the road to all road users.

Guide posts should be installed with red reflectors on the near side of the road and white reflectors on the far side of the road, as appropriate for LH or RH drive schemes. However, it should be borne in mind that reflective markers are only as good as the headlight efficiency which illuminates them, and they are also susceptible to reduced efficiency when obscured by dirt.

Solar powered road delineators are commonly used alternatives to reflective guide-posts and offer enhanced visual guidance, both at night and during bad weather. They provide conspicuous, legible reference points with the visibility (under ideal conditions) of several hundreds of metes, irrespective of headlight efficiency.

The type of guide posts and the installation method should prevent the possibility of reflectors being turned in the wrong direction, thus guide posts should either be driven into the ground or installed on timber or plastic pickets. Metal star pickets should not be used for the installation of guide posts due to the potential traffic hazard they represent.

Guide posts should be installed:

- Near the edge of formation at a consistent horizontal and vertical distance from the pavement edge
- At 50 m spacing on straight, flat sections of haul roads
- At 25 m spacing on curves and approaches to intersections
- As close to the edge of the carriageway as possible (but at a safe distance from road maintenance equipment)
- In pairs on both sides of the road
- With the height of the reflectors between 1.5 and 2.4 m from the road surface level.
- Green-type reflective markers should be installed exclusively at all (and only on) intersections, to aid recognition and differentiation of the junction.

Guide posts should be cleaned as part of an ongoing maintenance program and particularly after wet weather events and missing or damaged posts and markers should be replaced regularly. A maintenance and inspection program should also ensure that guide posts are visible during night-time driving conditions to both drivers of haul trucks and other vehicles.

Chapter 3

Structural design of haul roads

3.1 Background and orientation

Structural design, as with all other facets of haul roads design and operations, is focused on providing a safe and cost effective solution for the mining operations. This entails providing a road structure, often termed 'pavement', to protect the in-situ material from the heavy loads imparted by the mining trucks and to provide a vehicle friendly operating platform.

Figure 3.1 shows a typical cross-section of a haul road with the definition of the different layers. In the figure, two options are shown, namely:

i.   Typical in-pit road cross-section (LHS); and
ii. Typical ex-pit (surface) cross-section (RHS).

The existing material that needs to be trafficked is termed the 'in-situ' or 'sub-grade'. In cases where a fill or embankment is to be provided to achieve the desired geometric alignment, this material is considered to be the sub-grade. Generally there is little choice on the properties of the sub-grade, as they are defined by the mining operations. The softer the in-situ material is, the thicker the subsequent layer(s) must be to 'protect' the in-situ.

The pavement is that portion of the designed road structure placed above the sub-grade to support and provide the running surface for traffic. It must provide strength and protection as well as a surface with desired functional properties of acceptable ride quality with adequate skid resistance and minimal dust. In general highway engineering the pavement structure would consist of a sub-base, base and wearing course as illustrated. Since selected waste rock or overburden material is typically used for the structural layers, together with the economy of constructing a single layer rather than multiple structural layers, a mine haul road would only consist of a base rather than a sub-base and base. The upper-most wearing course will provide the desired functional properties and if designed correctly, the structure as a whole will provide an adequate response to the applied wheel loads over the design life and design traffic volume of the road.

A design based on and built according to the appropriate structural design specifications would accommodate various types of sub-grade or in-situ material and indicate how to 'cover' or place layerworks above them for adequate 'protection' as shown in Figure 3.2 to prevent premature failure such as illustrated in Figure 3.3.

Poor protection or too little 'cover' means that the sub-grade (or in-situ) will eventually deform and displace under the wheel loads of the trucks and the road will become unacceptably uneven, potholed and rutted. Because this layer is at the bottom of the road structure, it is expensive to repair when problems arise. This is termed 'structural failure' and Figure 3.3

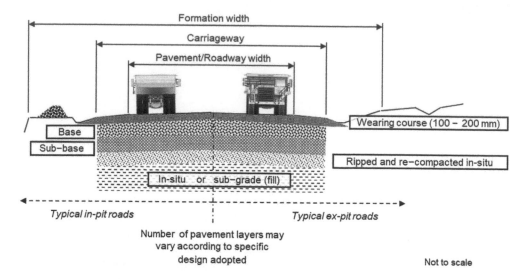

Figure 3.1 Cross-section of road structure describing the layers.

Figure 3.2 A road designed and built following an appropriate structural design specification showing no signs of structural failure – even after several years of high traffic volume operation.

Figure 3.3 Example of haul road structural failure due to inadequate cover above weak overburden material.

shows a typical example of a structural failure in a haul road, where the weak in-situ has deformed and displaced under the action of truck wheel loads. In this instance, either structural design (or layerworks 'cover' above in-situ) and/or the compaction of in-situ and/or layerworks were deficient. Rutting and large depressions of 0.4 m depth are seen. A distinction is made between this type of failure, and the more general 'functional' failure, in which the road does not perform its primary function of a safe, economical and vehicle friendly ride. This concept is explored further in Chapter 4. Suffice to say that a functional failure will always result where a structural failure is seen – even the best selection of road wearing course cannot perform its intended function without adequate support in the form of a suitable structural design.

The selection of the combination of layer materials and thickness to support the traffic loads during a design period is known as pavement or structural design. Structural design can be either an empirical or a mechanistic method. As shown in Figure 3.4 the empirical design is developed by analysing combinations of factors such as traffic, climate and material properties and obtaining responses to these combinations. The responses are then analysed statistically and a design method is developed. Such a method is dependent on the range of combinations that were tested. New materials with properties that were not evaluated cannot be used in the design. The CBR structural design method suffers from these limitations.

To overcome the limitations of an empirical method, a mechanistic or analytical method is used. In this method the variables are characterised in terms of fundamental properties such as stiffness and Poisson's ratio, and the stresses and strains under the applied load are calculated

Empirical Pavement Design

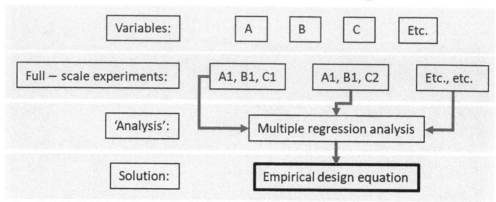

Mechanistic Pavement Design

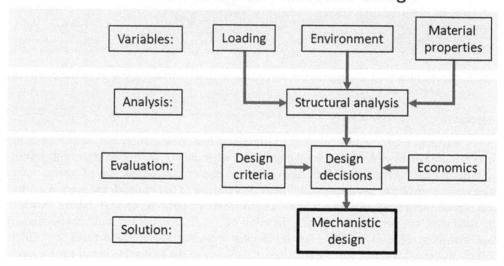

Figure 3.4 Illustration of empirical and mechanistic structural design methods.

in a structural analysis and used in the structural design. Although the mechanistic method has not been perfected it still has an empirical component as will be described in Section 3.3.

3.2 The CBR structural design method

3.2.1 Origin and background

During the late 1920s the California Division of Highways encountered severe distress on their roads in the form of rutting, which was as result of shear failure. At that time there was no formal structural design method, and the analysis was conducted with the aim of developing a design method. There was no material test method that could analyse the shear resistance quickly and cheaply. This led to the development of the California Bearing Ratio (CBR) test.

Samples of material were compacted in a mold (152 mm diameter × 127 mm high) in the laboratory with a given energy, and since the worst distresses on the roads were found to be during the wet season, the compacted samples were soaked for 4 days. After soaking, the sample was penetrated with a standard piston at a given loading rate, and the resistance at 0.1″ (2.54 mm) was measured. Figure 3.5 shows a schematic of the test procedure as well as the soil cone that forms under the piston, which then results in a shear failure. This is similar to the analysis of structural footings. Often engineers are concerned that this test method is unrelated to the mode of failure of a pavement, but this is not correct as demonstrated in the figure. After penetration and correction for seating of the piston, the applied resistance is divided by the resistance of what was considered at the time as the average best crushed stone, and expressed as a percentage. This method is a standard method in most countries, although the American ASTM or AASHTO methods are the standard as used in the CBR thickness design method. In all but arid and semi-arid environments, the CBR value adopted in the design should be based on a soaked CBR test, commonly for 4 days, but up to 10 days for roads in high rainfall or tropical climates.

A limitation of this method is that because of mold edge effects the maximum particle size that can be used is 19 mm. Larger particles would provide a greater resistance, but these are either discarded, or crushed. Both techniques result in changes to the grading, but in the absence of a satisfactory solution, the results are accepted acknowledging that the support could be greater. Note that the CBR was calculated with the maximum value of 100% reflecting the best available crushed stone. Values greater than 100% therefore have no meaning as better quality materials were not available and were not tested. Extrapolations therefore have no engineering meaning except that the material is better than the 'best'.

After the investigations a CBR-based pavement thickness design method was developed.

3.2.2 CBR thickness design method

The CBR thickness design method determines the cover or protection thickness of a material for a given support or CBR bearing capacity. This method supports a fundamental requirement

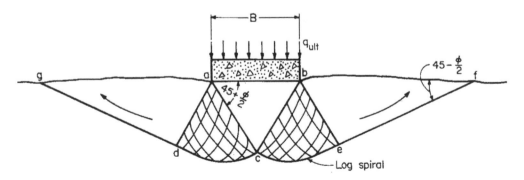

Figure 3.5 Bearing capacity failure under a surface footing or tyre, also applicable to the CBR test, where B is the piston.

Source: after Bowles, 1984.

that a certain protection is required over a weaker layer. Unfortunately the method does not take into account the strength or support of the overlaying materials, nor does it consider other non-natural materials such as cemented or bituminous materials.

Furthermore, considering the standard American pavement design approach at that time, the cover thickness curves, such as those shown in Figure 3.6, show that the curves stop short at a thickness of 75 mm (3″). This was because it was accepted that a minimum thickness of 75 mm of asphalt would be required as surfacing. For heavier traffic conditions, the asphalt thickness is increased. This aspect is normally ignored when the method is used for pavements without an asphalt surfacing such as unpaved mine haul roads.

The cover thickness curve is used to determine the protection for each successive layer based on the CBR of a layer and bearing in mind practical layer thicknesses and cost of materials and construction. It is for example not practical to build a pavement with a single material with a CBR of say 80, which is a good but expensive material. In such a case, progressively stronger materials are used in the upper layers.

In the original research, three traffic categories were identified, namely light, medium and heavy. This was the road classification used in the 1920s. Subsequently wheel loads of 31 kN (7000 lb), 40 kN (9000 lb) and 53 kN (12,000 lb) were associated with these qualitative terms, an action which could be questioned.

During the early 1940s the US Corps of Engineers needed to construct airfields on the Pacific islands. At the time no airfield structural design procedure was available, and the CBR highway structural design method was adapted for airfield pavements. The adaptation was to

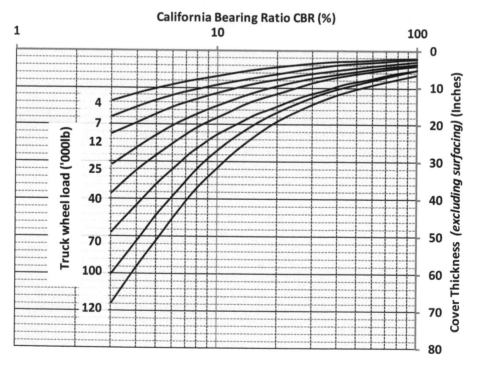

Figure 3.6 Kaufman and Ault (1977) haul road thickness design chart.

relate the shear stresses in a pavement to wheel load. This formed the basis for the CBR thickness design for surface haul roads developed by Kaufman and Ault (1977), and the design chart (in Imperial units) is shown in Figure 3.6.

Several refinements to overcome limitations were made after WWII. A relationship between flexible pavement thickness to wheel load and tyre pressure was developed. Further studies allowed the incorporation of a thickness adjustment to account for traffic repetitions. Despite the empirical origins of the technique, Turnbull and Ahlvin (1957) derived a mathematical approach to the calculation of cover requirements using the CBR method. This approach is adopted for the calculation of cover requirements over in-situ material, as predicted by the CBR design method for ultra-heavy axles. For the complete mathematical treatise, readers are referred to Turnbull and Ahlvin (1957), Ahlvin *et al.* (1971) and Otte (1979).

A further limitation that was identified was that the design method only considered single wheels whereas dual wheel rear axle configurations (common on most mine rear dump trucks), would apply greater stresses with depth than a single wheel, which could not be handled. Kaufman and Ault (1977) increased the single wheel load of a dual configuration by 20% to achieve an equivalent single wheel load (ESWL), assuming an equal contact area.

3.2.3 Example of application

A reworked metric version of the latest thickness design chart is shown in Figure 3.7. It also facilitates application to a specific vehicle type, but the influence of dual tyres needs to be incorporated by increasing the wheel load by 20%, as with the original Kaufman and Ault approach. Note that the original chart in Figure 3.6 required at least 3″ of cover for a CBR

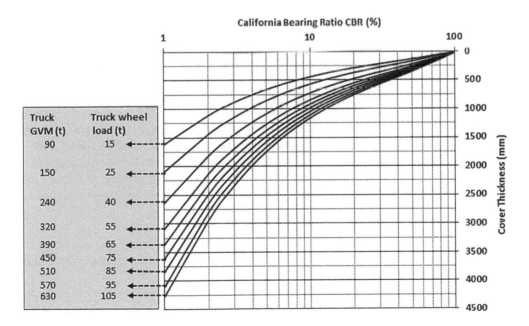

Figure 3.7 CBR thickness design chart.

Source: reworked from Kaufman and Ault, 1977.

of 100. In the reworked version there is no such requirement and all curves have a zero thickness requirement at 100 CBR.

Conventionally, the role of the wearing course is not considered in the CBR cover-curve design approach. However, with mine haul roads, due to the recommended design thickness of this layer (200 mm for long life roads and 100 mm for roads with a life less than 1 year) and the regular maintenance the roads are subjected to, the omission of this layer may be a somewhat conservative assumption.

The following equation can be used to estimate the layer thickness [Z_{CBR} (m)] required for a material of California Bearing Ratio (CBR %):

$$Z_{CBR} = \frac{9.81 t_w}{P} \left[0.104 + 0.331 e^{(-0.0287 t_w)} \right] \left[\left(2 \times 10^{-5} \right)^{\left(\frac{CBR}{P} \right)} \right] \left[\left(\frac{CBR}{P} \right)^{-\left(0.415 + P \times 10^{-4} \right)} \right] \qquad \text{Equation 3.1}$$

Where t_w is the truck wheel load (metric tonnes), P is tyre pressure (kPa) and CBR is the California Bearing Ratio of the layer material (%).

Originally (Kaufman and Ault, 1977) the truck wheel load was increased by 20% to replicate the effects of the increased stresses generated by a rear dual-wheel axle which occur deeper in a road layer – the concept of Equivalent Single Wheel Load (ESWL). From a more formal approach to ESWL following Foster and Ahlvin and a semi-rational ESWL approximation, the following equation can be used to estimate the cover Z_{ESWL} (m) more reliably as:

$$Z_{ESWL} = Z_{CBR} + \left[0.184 + \left(0.086 CBR + \frac{17.76 CBR}{t_w} \right) \right]^{-1} \qquad \text{Equation 3.2}$$

Note that when applying these formulae to a design determination, due to the estimation characteristics of the formulae, a more realistic *final* layer thickness is found by subtraction of the sum of preceding layers from total cover requirements. The final design would then incorporate the wearing course layer above the total cover thickness determined from the previous equations or the chart shown in Figure 3.7.

As a demonstration of the thickness design chart, a CAT793D rear dump truck with GVM 384 t, 67% carried on the rear single axle with dual tyres (SADT) and a tyre pressure of 700 kPa will be used. From previous studies it was found that the in-situ material is a silty-clay with a CBR of 5%. The only other material that is available from the mining operations is a sand with a CBR of 15%. An on-site crusher produces a graded crushed stone with a CBR of 80%. The different steps required to determine total cover and individual layer thicknesses are shown in Figure 3.8.

Step 1: Enter the chart with the in-situ CBR of 5% and move vertically until the increased ESWL of 384 × 0.67 × 1.2/4 = 77.2 t curve. Read the cover thickness on the vertical axis, which is 1750 mm.

Step 2: Enter the chart with the CBR of the sand of 15%, and read off the cover thickness from the vertical axis. This is 750 mm. The thickness of the sand layer is thus 1750–750 = 1000 mm.

Step 3: The cover thickness over the sand is also the thickness of the graded crushed stone which has a CBR of 80% and is thus 750 mm.

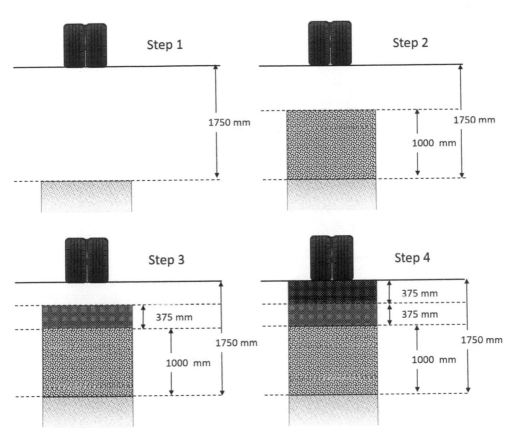

Figure 3.8 Determination of layer thickness for a laden CAT793D haul truck.

Figure 3.9 shows this design process superimposed on the CBR design chart.

The same process can be repeated, using Equation 3.1 and Equation 3.2. Since Equation 3.2 determines the cover required for an ESWL using a more formal approach to ESWL (following Boyd and Foster, 1950), a rear wheel load for the CAT793D of 77.2 t is used as variable t_w and 700 kPa for variable P. Table 3.1 summarises the results which are similar to those determined graphically (with the 20% increase in wheel loads), but incorporating a more reliable estimate of Z_{ESWL}. However, as will be seen later when comparing structural designs derived from both the CBR and the mechanistic approaches, even with a refined estimate of cover from the ESWL, the under-design of the CBR cover will be apparent.

This has completed the cover thickness design. Although the design chart provides a solution for in-situ material with a CBR of 1, care should be taken with materials with a CBR of 3 or less, as these could be highly plastic clays. Such materials are unstable, will not support construction traffic and will require special drainage attention.

As has been shown in the previous example, road cover thickness above a material with a particular CBR is determined as a function of applied wheel load and the CBR of the material itself. The same technique can be used for successive layers – the only requirement being

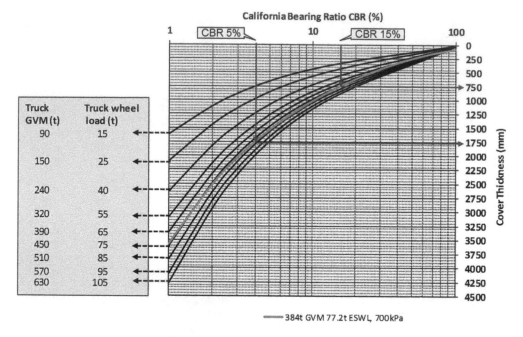

Figure 3.9 CBR cover-curve-based solution for example discussed.

Table 3.1 Results of CBR cover curve calculation using Equations 3.1 and 3.2 for a two-layer road.

CAT793D GVM (tonnes)	384
Wheel load (tonnes)	64.3
ESWL *estimate* (tonnes)	77.2
Tyre pressure (kPa)	700

Layer	CBR	Single Wheel			ESWL		
			Cover	Layer thickness		Cover	Layer thickness
In-situ	5	Z_{CBR}	1.432 m		Z_{ESWL}	1.934 m	
Sand layer	15	Z_{CBR}	0.720 m	0.712 m	Z_{ESWL}	0.898 m	1.035 m
Crusher run layer	≥80			0.720 m			0.898 m

that successive layers must be of higher CBR than the preceding layer and any layer to layer increase in CBR limited to about 2–2.5 times the preceding CBR value.

In a similar example, with additional lower- and upper sub-base layering, consider the 'cover' required for the same truck as previously. As shown in Table 3.2, with an in-situ sub-grade CBR of 5%, the cover (pavement thickness) required is 1934 mm. If the sub-grade were ripped and recompacted to form a lower sub-base layer of CBR 16%, pavement cover required

Table 3.2 Results of CBR cover curve calculation using Equations 3.1 and 3.2 for a four-layer road.

CAT793D GVM (tonnes)	384
Wheel load (tonnes)	64.3
ESWL *estimate* (tonnes)	77.2
Tyre pressure (kPa)	700

Layer	CBR	Single Wheel			ESWL		
			Cover	Layer thickness		Cover	Layer thickness
In-situ	5	Z_{CBR}	1.432 m		Z_{ESWL}	1.934 m	
Sand layer	16	Z_{CBR}	0.687 m	0.745 m	Z_{ESWL}	0.855 m	1.079 m
Crusher run layer	28	Z_{CBR}	0.435 m	0.252 m	Z_{ESWL}	0.532 m	0.323 m
Lower Base	60	Z_{CBR}	0.183 m	0.252 m	Z_{ESWL}	0.229 m	0.303 m
Upper Base/Wearing Course	≥80			0.183 m			0.229 m

(above ripped and recompacted sub-grade) is now 855 mm so (1934–855) = 1079 mm layer thickness is required (most likely as three lifts if compacted with vibratory rollers). Placing an upper sub-base of CBR 28% results in a cover requirement of 532 mm so (855–532) = 323 mm layer thickness, following which a CBR = 60% base is placed, which requires cover of 229 mm and thus 303 mm layer thickness. Ideally, a yet harder material is required, CBR > 80% and would for design purposes be specified to 229 mm (practically rounded to 250 mm) depth.

For the CBR thickness design method material characterisation is in terms of the support determined by the CBR test, as described in Section 3.2.1. Remember that the maximum size of particles is 19 mm, and coarse material or dump rock cannot be evaluated.

For a new haul road samples may be taken and tested in a soils laboratory. In the mining environment production often does not allow the liberty of waiting for laboratory test results. An alternative test is the Dynamic Cone Penetrometer (DCP), which is shown in Figure 3.10.

3.2.4 Material characterisation and laboratory tests

The DCP measures the resistance of the soil at the in-situ moisture content and density. The hammer is dropped over a distance of 575 mm under its own mass, and the penetration of the cone into the soil is measured on the 1 m tape measure. Normally the penetration is recorded for every five blows of the hammer. The resultant penetration is converted to the CBR from the correlation curve shown in Figure 3.11. Note that the correlation curve is only valid for the dimensions of the DCP shown. Other DCP models with different dimensions exist which would not give the correct CBR. If the in-situ measurements are taken on a material which is moist but may become saturated because of poor drainage, then the CBR is an overestimate, and should be reduced.

The DCP can also be used to determine the CBR profile with depth on an existing haul road. Penetration is usually taken to a depth of 800 mm. The actual profile with depth may then be compared with the design, as shown in Figure 3.12. Layers weaker than the design

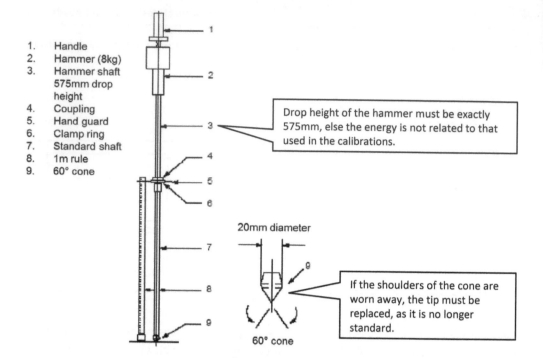

1. Handle
2. Hammer (8kg)
3. Hammer shaft
   575mm drop
   height
4. Coupling
5. Hand guard
6. Clamp ring
7. Standard shaft
8. 1m rule
9. 60° cone

Drop height of the hammer must be exactly 575mm, else the energy is not related to that used in the calibrations.

20mm diameter

If the shoulders of the cone are worn away, the tip must be replaced, as it is no longer standard.

60° cone

Figure 3.10 Schematic of the Dynamic Cone Penetrometer (DCP).

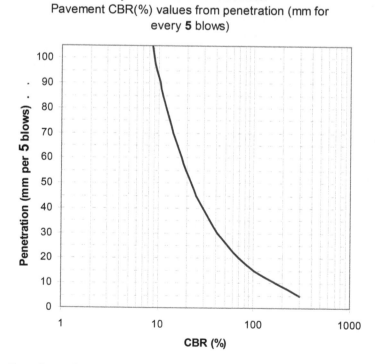

Dynamic Cone Penetrometer
Pavement CBR(%) values from penetration (mm for every **5** blows)

Figure 3.11 Correlation between DCP penetration per five blows and CBR.

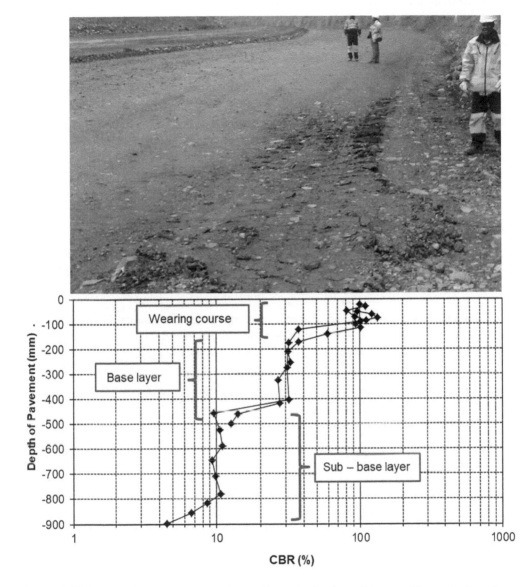

Figure 3.12 Severe distress seen on the surface of a haul road (upper illustration) and the corresponding evaluation of the DCP derived CBR profile of the structure in that location, showing a weak and thin base layer which accounts for the distress.

are easily recognised, and additional protection or replacement may be required. It is also an easy way to determine weak layers that may be responsible for poor performance. The DCP is not able to penetrate dump rock or coarse rock layers, and when refusal has been reached the test must be stopped to avoid damaging the DCP.

The same procedure is used to determine the properties of an existing haul road for application in the mechanistic method, as will be discussed in the next section. This method is also used to evaluate haul road pavements when larger haul trucks are to be introduced.

3.2.5 Limitations of the CBR method

In the presentation of the CBR thickness design several limitations of the method, as well as in the CBR test, were mentioned. Notwithstanding that the CBR thickness design has been widely used because of ease of application, it is useful to list these limitations, as follows:

- Coarse material of particle size greater than 19 mm cannot be tested. This would lead to an underestimate of the bearing capacity.
- The method does not consider the inherent strains and resultant deflections in weak layers such as the sand layer in the CBR pavement example. High surface deflections increase rolling resistance.
- The CBR method does not take into account the properties of the surface material.
- The CBR method was originally designed for paved roads and surfaces for airfields. Therefore the method is less applicable for unpaved roads, especially haul roads which experience much different wheel geometry and construction materials despite adaptations.
- The empirical design curves were not developed for the high axle loads generated by large haul trucks and simple extrapolation of existing CBR design curves can lead to errors of under design, or even over design.

Although the CBR thickness design method was a major contribution in understanding pavement design, particularly the need to provide protection, more advanced methods are now available for pavement design as will be discussed in the next section.

3.3 Mechanistic-empirical design method

3.3.1 Origin of method – Boussinesq theory

In 1885 Boussinesq developed equations for calculating the stresses, strains and deflections under a point load at the surface for a homogeneous, isotropic, elastic and semi-infinite layer (an infinite depth). Interestingly the vertical stress at any depth of a single material due to a point load at the surface is dependent on the depth and radial position and is independent of the properties of the transmitting medium (Yoder and Witczak, 1975).

The original equations had little application to reality, as wheel loads are typically circular or elliptical, and pavement structures consist of several layers. Subsequently the US Corps of Engineers developed solutions for a Poisson's ratio of initially 0.5, and later for any value of Poisson's ratio. This allowed the calculation of stress, strain and deflection in a pavement consisting of an upper layer and subgrade, by making simplifying assumptions. Burmister (Yoder and Witczak, 1975) extended the solution to two layers by using a number of charts with factors. It became messy to apply this as a design procedure, and impractical for more layers.

Computers and computing technology had in the meantime advanced significantly and in the early 1960s the first computer generated solutions for stress, strain and deflection using a linear elastic multilayer analysis became available. This allowed for routine calculation of these parameters and formed the basis for current mechanistic design. Linear elastic materials have a linear stress-strain loading curve, as opposed to non-linear materials that have a curved response. There are computer packages that allow non-linear analysis, but since determining the non-linear input parameters require sophisticated laboratory facilities, this method is generally not used on mines. Other packages allow the use of different stiffness values in the vertical and horizontal directions, termed anisotropy. Again, obtaining the relevant input parameters require sophisticated laboratory facilities. This level of sophistication may not be necessary as in practice it was found that linear elastic multilayer analyses give pavements that provide good performance. Finite element packages are also available but invariably the creation of the input is a tedious process, and the outcome is similar to that of the simpler linear elastic multilayer analysis.

A number of public shareware computer packages as well as commercially developed packages are available on the market. Most of the packages use the same computing engine, but the differences are the input and output format. As part of the output development additional processing may take place, and users are encouraged to ensure that these refinements provide the desired outcomes. One of the earliest shareware packages is ELSYM5A (FHWA, 1985), which analyses up to five layers, and which is primitive as it still uses the original Fortran code. There are commercial adaptations such as MePADS to facilitate input and outputs. Position in ELSYM5A is in Cartesian coordinates. A refinement is CHEV15, which analyses 15 layers, a facility which may not be necessary on mine haul roads. Another shareware package is KENLAYER, which is issued together with a textbook produced by Huang (1993). The Australian CIRCLY package is commercially available and allows analysis of anisotropy, which again is not necessary on haul roads because of difficulty in obtaining the input parameters. However, CIRCLY comes with a front-end specifically designed for haul road applications, in which vehicles and pavement layerworks can quickly be defined and solved.

3.3.2 Critical parameters for analysis and input measurements

For a mechanistic analysis the loads, typically wheel loads (P), and the layer properties such as thickness (h), elastic modulus or stiffness (E) and Poisson's ratio (μ), as shown in Figure 3.13, are required. The critical parameters for fine-grained, coarse-grained or dump rock materials are the vertical compressive strain at the top of the respective layers. The vertical compressive strain is normally expressed as microstrain ($\mu\varepsilon$) or strain times 10^{-6}. On highways where other materials are used, the critical parameter of cement stabilised or asphalt layers is the horizontal tensile strain at the bottom of the layer, and for highly compacted crushed stone base layers it is the shear stress factor of safety in the middle of the layer. On mine haul roads cemented layers are normally not used, and because of the large tyre size the stress conditions in the wearing course are not critical.

The critical points for analysis are midway between the two wheels of the dual wheel combination (for influences at depth) and directly under the wheel for the upper pavement, as shown in Figure 3.14. The wheel loads must be positioned to take account of superposition of stresses in the vertical plane as shown schematically in Figure 3.15. The positions of analysis are shown geometrically in Figure 3.16. This is an exhaustive analysis for a conventional single front axle single tyre (SAST) and single (rear) axle dual tyre (SADT) rear dump truck, where the critical positions are determined both from consideration of the truck wheel base and dual tyre geometries, coupled with the GVM split between SAST and SADT for laden conditions. It is generally found that the rear axle dual wheel combination on one side gives the most critical values; including wheels on the other side of the axle or the front axle has virtually no influence. A rigorous analysis will dispel any doubts.

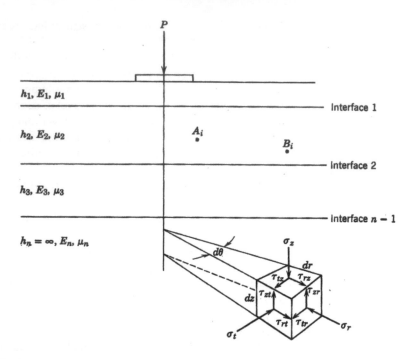

Figure 3.13 Generalised multilayer elastic system.

Source: Yoder and Witczak, 1975.

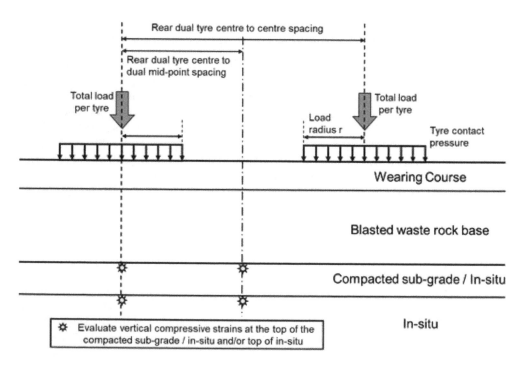

Figure 3.14 Schematic representation of data and solution requirements for a linear elastic multi-layer analysis of a haul road (in this case, incorporating a compacted in-situ layer).

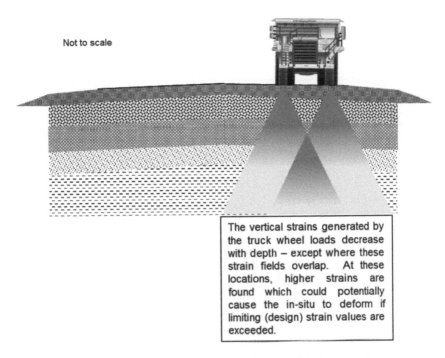

Not to scale

The vertical strains generated by the truck wheel loads decrease with depth – except where these strain fields overlap.   At these locations, higher strains are found which could potentially cause the in-situ to deform if limiting (design) strain values are exceeded.

Figure 3.15 Schematic representation of stresses generated below truck.

OEM data should be consulted to determine the load geometry, generally based on a single dual assembly for the appropriate laden condition of the truck. The dual wheel centre to centre spacing [S_d (m)] is required, as defined in Figure 3.17, following Equation 3.3:

S_d = (*Overall rear dual tyre width – Centreline rear dual tyre width*)
   – Tyre width   Equation 3.3

In the absence of specific geometric details, S_d (m) can be estimated using the tyre width and a dual clearance of approximately 0.3–0.4 m. Alternatively, the following Equation 3.4 applies to most large rear dump mining trucks (RDTs):

$S_d = 0.08\,\text{GVM}^{0.488}$ Equation 3.4

Wheel load is the important input with regards to strains on the subgrade. Often, manufacturers will accommodate a 10:10:20 overload rule which implies that no more than 10% of truck loads should exceed 110% of payload and none may exceed 120%. Thus, the OEM rated payload may be increased by 10% to recalculate a more realistic value for GVM and thus wheel loads. Additionally, but dependent on the location of each truck's centre of gravity, there will be a load transfer on a ramp in which either the front or rear axle will carry an increased proportion of the GVM. As a general rule, axle load increases with about 1.6 × grade (%) of the road, whilst camber/crown would transfer 1.15 × crown/camber (%) to one

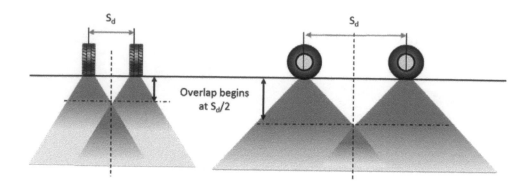

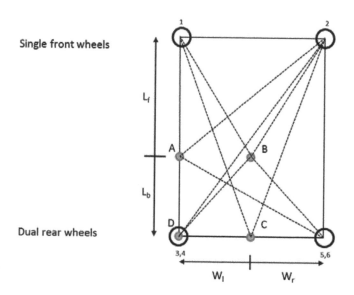

Figure 3.16 Critical positions for a fully loaded haul truck.

Source: Thompson, 1996.

or other side of the truck. Thus a truck climbing a 5% ramp where a 2% crossfall is applied would experience a 10.3% increase on the calculated axle load. This would then be apportioned according to the specific axle under consideration.

Table 3.3 summarises a wheel load determination, in this case for a CAT793D on a flat haul (0% grade) and 2% camber. Maximum design wheel loads for the two rear tyres in the dual assembly is 640.7 kN at OEM GVM and 677.1 kN when the 10% payload increase is considered. For the same truck on a 10% ramp with 2% camber applied and carrying the extra 10% payload, this increases to 783.8kN (Table 3.4).

Tyre pressures are required; based on OEM specifications, these are generally 700–900 kPa. The tyre pressure is not critical as it influences the wearing course, which is not

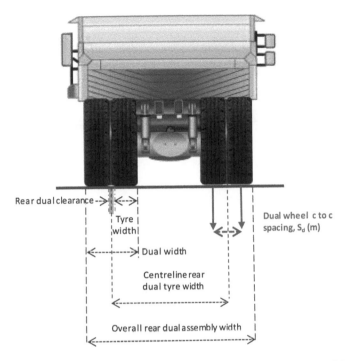

Figure 3.17 Definition of dual wheel centre to centre spacing in relation to OEM dimensions.

analysed separately, but it does dictate the contact pressure between the tyre and the structure below. In the majority of design problems, a circular tyre imprint is assumed and the contact radius (*a* in metres) can be determined from Equation 3.5:

$$a = \sqrt{\frac{W_t}{P\pi}}$$

Equation 3.5

Where
W_t = Weight on a single tyre (kN)
P = Tyre pressure (kPa)

The strains induced in a pavement are a function of the effective elastic modulus values assigned to each layer material in the structure. The pavement materials are characterised by the elastic modulus, also known as stiffness, and the Poisson's ratio. Both parameters are measured in a repeated load large triaxial cell (300 mm high and 150 mm diameter samples) normally in a specialised soils laboratory. This level of sophistication is not necessary in the mining environment, and a generalised indication of these input parameters will suffice.

In conjunction with the effective elastic modulus, the Poisson's ratio is also required, which is an indication of how vertical deformation converts to horizontal stresses and strains. When there is minimal transfer the Poisson's ratio would be zero, and for a full conversion the ratio is 0.5. Concrete for example has a Poisson's ratio of 0.2, and gravel and granular materials have a value of between 0.35 and 0.44. The latter is typically used in analyses.

Table 3.3 Determination of truck wheel loads, 10% payload overload, flat haul and 2% camber.

Application Specifications				
Road Grade (%) (+ve against the load)	0			
Road Crown/crossfall (%)	2			

Haul Truck Specifications				
Haul Truck Type:	**CAT793D**			
EVM	166			
Payload	218			
Payload 10:10:20 Applied (Y or N)	Y			
GVM (OEM Specification)	384			
GVM (EVM + 10% Payload Overload)	405.8			
Overal width across rear axle (m)	7.605			
Centreline rear dual tyre assembly width (m)	4.963			
Dual tyres centre to centre spacing (m)	**1.626**			

Axle Groups				
	Front Axle Group		**Rear Axle Group**	
Axle Group Type	SAST		SADT	

Tyre data				
Tyre type (width) (")	40		40	
Operating tyre pressure (kPa)	700		700	

Wheel Load Options				
Weight Distribution (%)	33		67	
	Wheel Load (kN)	Contact radius (m)	Wheel Load (kN)	Contact radius (m)
At OEM GVM (kN)	621.56	0.532	630.98	0.536
plus load transfer (kN)	626.28	0.534	640.70	0.540
At OEM GVM + 10% payload overload (kN)	656.85	0.547	666.80	0.551
plus load transfer (kN)	661.83	0.549	677.08	0.555

Besides obtaining the elastic modulus from repeated load triaxial testing, this value may also be obtained by in-situ measurements of deflection with depth using the multi-depth deflectometer (MDD) (de Beer *et al.*, 1989). The MDD is a series of linear variable differential transformers (LVDT) fixed to a central core which is placed in a 40 mm core hole and anchored at the desired depth as shown in Figure 3.18. The LVDTs are placed at the depth of the layer interfaces and locked into the layer such that movements under load are measured. By performing a trial and error analysis in assuming stiffness values with a linear multilayer package such as ELSYM5, the calculated deflections at depth are matched with the measured deflections. This procedure was used to confirm the elastic modulus of typical in-situ materials

Table 3.4 Determination of truck wheel loads, 10% payload overload, 10% ramp (uphill) and 2% camber.

Application Specifications				
Road Grade (%) (+ve against the load)	10			
Road Crown/crossfall (%)	2			

Haul Truck Specifications				
Haul Truck Type:	**CAT793D**			
EVM	166			
Payload	218			
Payload 10:10:20 Applied (Y or N)	Y			
GVM (OEM Specification)	384			
GVM (EVM + 10% Payload Overload)	405.8			
Overal width across rear axle (m)	7.605			
Centreline rear dual tyre assembly width (m)	4.963			
Dual tyres centre to centre spacing (m)	**1.626**			

Axle Groups				
	Front Axle Group		**Rear Axle Group**	
Axle Group Type	SAST		SADT	

Tyre data				
Tyre type (width) (")	40		40	
Operating tyre pressure (kPa)	700		700	

Wheel Load Options				
Weight Distribution (%)	33		67	
	Wheel Load (kN)	**Contact radius (m)**	**Wheel Load (kN)**	**Contact radius (m)**
At OEM GVM (kN)	621.56	0.532	630.98	0.536
plus load transfer (kN)	526.83	0.489	741.66	0.581
At OEM GVM + 10% payload overload (kN)	656.85	0.547	666.80	0.551
plus load transfer (kN)	556.74	0.503	783.76	0.597

as shown in Table 3.5, as well as for the wearing course material and the dump rock layer (Thompson and Visser, 1996).

Alternatively, effective elastic modulus (E) estimates for subgrade layer materials can be based on AustRoads, Angell (1988) or (slightly more conservatively) the Powell *et al.* (1984) model and California Bearing Ratio (CBR) relationship, as shown in comparison in Figure 3.19.

- Angell (1988) (CBR < 15) $E = 21.1CBR^{0.64}$
- Angell (1988) (CBR > 15) $E = 19CBR^{0.68}$

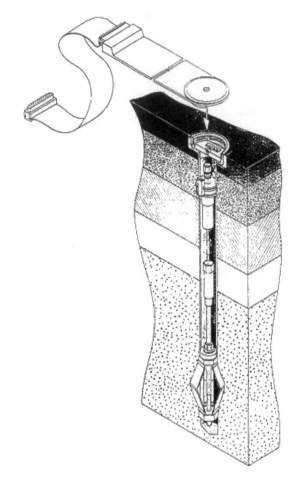

Figure 3.18 Illustration of the multi-depth deflectometer.
Source: de Beer *et al.*, 1989.

Table 3.5 Typical elastic modulus of lower pavement materials.

CBR (%) of in-situ material	Effective elastic modulus (MPa)	
	Wet state	*Dry state*
CBR ≥ 25	105	135
24 ≥ CBR ≥ 15	85	135
14 ≥ CBR ≥ 10	65	120
9 ≥ CBR ≥ 7	55	95
6 ≥ CBR ≥ 3	45	65

Source: Thompson and Visser, 1996.

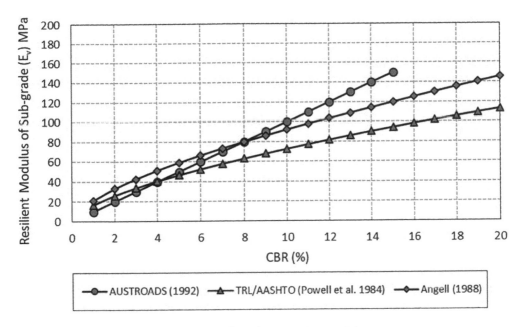

Figure 3.19 Relationships to predict subgrade or in-situ modulus.

- AustRoads (1992) (CBR < 15) $E = 10CBR$
- TRL/AASHTO (Powell *et al.*, 1984) $E = 17.63CBR^{0.64}$

AustRoads (2009a) also propose typical subgrade strength values based on classification following USCS, as shown in Table 3.6.

Other estimation techniques, typically based on correlations to gradation and Atterburg limits, for example the VicRoads method (specified in Materials Engineering Branch; Engineering Road Note 9, MainRoads WA Australia, 2013) can be used but may require calibration to local conditions since the range of sub-grade materials' CBR values modelled by the estimation equations will vary from location to location. Equation 3.6 and Equation 3.7 apply when $P_{2.36} > 75\%$ and PI > 6 under well-drained conditions and should not be used for projects north of latitude 26°S without specific investigation and verification:

$$log_{10}CBR = 1.668 - 0.00506P_{0.425} + 0.00186P_{0.075}$$
$$- LS\left(0.0168 + 0.000385P_{0.075}\right) \qquad \text{Equation 3.6}$$

$$log_{10}CBR = 1.886 - 0.00372P_{2.36} - 0.0045P_{0.425}$$
$$+ \left(\frac{P_{0.075}}{P_{0.425}}\right)\left(5.15 - 0.0456\left(\frac{P_{0.075}}{P_{0.425}}\right)\right)10^{-3} - 0.0143PI \qquad \text{Equation 3.7}$$

Table 3.6 Estimation of subgrade CBR using USCS.

Description of Sub-grade		Typical CBR values (%)	
Material	USCS Symbol	Excellent to good drainage	Fair to poor drainage
High plasticity clay	CH	4	2–3
Silt	ML	5	2
Silty-clay/Sandy-clay	CL	5–6	3–4
Sand	SW/SP	10–18	10–18

The estimate of subgrade CBR is determined from Equation 3.8, where CBR_{min} refers to the lower of the two values from Equations 3.6 and 3.7 and CBR_{max} the larger value:

$$CBR = 0.25\left(3CBR_{min} + CBR_{max}\right)$$

Equation 3.8

Table 3.7 recommends modulus value correlations to USCS and AASHTO classification systems. To facilitate the choice of suitable modulus values for in-situ materials, the associated range of CBR values derived in the field from Dynamic Cone Penetrometer (DCP) probing are also given. Actual test results may prove the bearing ratios for a specific soil group to be considerably better than the low value depicted on the chart. However, it is prudent to adopt conservative estimates in most cases since the following:

- Subgrades, especially in tropical climates, can vary in their weathered state to become weaker in localised areas.
- It is difficult to ensure that the conditions associated with higher than anticipated CBR results will prevail unchanged for the full design life of a haul road.
- The risk associated with a high design CBR value is the very high cost of remediation if premature structural failure occurs.

With regards to the layer above the sub-grade or in-situ, in the original research in which the mechanistic approach to mine haul road design was developed (Thompson, 1996), the stiffness of this layer, when comprised selected blasted dump rock, was found to be 3000 MPa, which is similar to the initial stiffness of a cement stabilised gravel with about 3% cement. Obviously this material could not be tested in the triaxial cell. Where compaction is poor, and the material of marginal quality or lifts (layer thickness) excessive, this value should be reduced to 1500–2000 MPa. This would be more typical of a rubblised PCC base layer, to which a poorly compacted and only partially interlocked selected blasted waste rock layer would correspond.

Minimum specifications for the source material chosen for this dump rock layer include the following:

- Uniaxial compressive strength in excess of 50 MPa
- Not containing weathered rock, clay or soil with no more than approximately 20% minus 2 mm material and predominantly freely draining material when placed and compacted
- Low abrasive losses (e.g. Los Angeles abrasive index of no more than 30%)
- Have a minimum particle density of 2.0 t/m^3

Table 3.7 Effective elastic modulus correlations with the AASHTO and USCS soil classification systems.

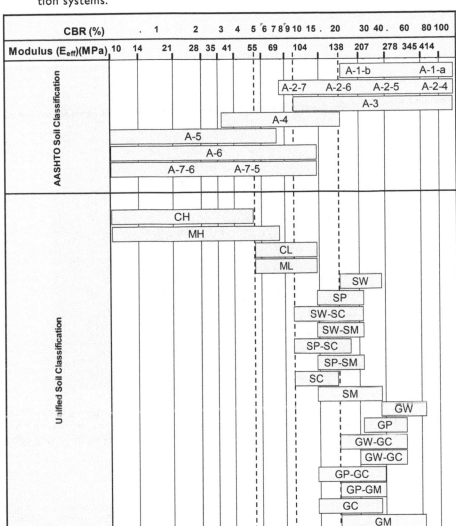

- The largest block size is ideally ⅔ lift thickness which equates to between 200 and 300 mm maximum for most haul road layer thicknesses. Larger rock is difficult to compact and forms a high spot in the layer surrounded by a ring of soft uncompacted material.

Figure 3.21 shows a typical source of selected blasted waste rock.

The wearing course, which should have a CBR of at least 60, and preferably 80, has a stiffness of 300–400 MPa when placed over a dump rock layer which has excellent support.

Figure 3.20 Use of a selected blasted dump rock layer placed directly above in-situ material and immediately below the wearing course is a characteristic of the mechanistic approach to mine haul road design. Here such a layer is seen following compaction and blinding.

Figure 3.21 Typical material for the selected blasted waste rock base layer.

Further specifications for the wearing course, relating also to its functional role, are discussed in Chapter 4.

Using a mechanistic design methodology, the layer specifications are based on limiting load-induced strains in the softer sub-grade/in-situ to below certain critical values. These values are associated with the category of road being designed, truck size, performance requirements and road operating life. The higher the traffic wheel loads and volumes (kt/day) and the longer the operating life and associated performance requirements are of the road, the lower is this critical strain value. This data is then used to determine the thickness of the selected blasted dump rock layer to be placed on top of the in-situ, sub-grade or fill such that the road will perform satisfactorily over its design life.

Typical limiting sub-grade vertical compressive strain values from various sources are summarised in Figure 3.22, together with the Thompson and Visser (1996) recommended values. The horizontal scale of this graph is the product of the daily traffic in kiloton and the performance index, which relates to the expected performance of the road. The anticipated life of the road would also dictate the limiting strain criteria, as long-term roads would require better performance. The classification of the road network would assist with selection of the limiting strain and Equation 3.9 can be used to estimate the limiting strain up to a maximum of 200 (product of daily traffic volume as mass (kt) and road performance index RPI). Although the line could be extrapolated, it has been found that for ultra-high tonnage, high traffic volume, long-term roads, the lowest strain value would asymptote at 800 µℇ:

$$\varepsilon_{Lim} = 4035e^{-0.007kt.RPI} \qquad\qquad \text{Equation 3.9}$$

The choice of 2000 as the limiting vertical compressive strain was derived from the South African mechanistic design procedure. Figure 3.23 shows the transfer function that relates rutting to number of load applications as a result of the applied vertical compressive strain. Similar curves are used in most mechanistic design methods. The rutting occurs under a 80 kN axle load, which is a typical design load. Based on the applied load and calculated compressive strain, it was argued that these curves could then be applied to rutting under any load. On mine haul roads 10,000 load repetitions was considered to be a practical input value, and 2000 µℇ would produce a 20 mm rut. Since haul roads would normally be frequently maintained by using a motor-grader, such ruts could be easily corrected.

3.3.3 Analytical methodology

To determine the layer response to an applied load, a layered elastic model is used to represent the various haul road layers in the design. As discussed previously, software is available which can be used to solve multi-layer problems in road design, including ELSYM5, MePADS and CIRCLY6. Irrespective of the solution software used, the theoretical approach is similar and is summarised in Figure 3.24.

Typical results would include an assessment of modelled versus critical strains at the sub-grade boundary, together with an indication of maximum surface deflections, associated with the proposed/existing design and truck types, as shown in Table 3.8. In this analysis, a Category II road carrying 50 kt/day with an RPI of 2 was used to determine the limiting strain value of 2000 µe from Figure 3.24.

It is not anticipated that most mining engineering professionals would perform this type of mechanistic structural analysis on a daily basis. This is a specialised activity and would

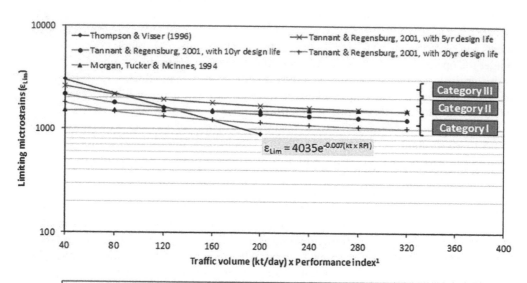

The equation shown in the chart:

$$\varepsilon_{Lim} = 4035e^{-0.007(kt \times RPI)}$$

Chart legend:
- Thompson & Visser (1996)
- Tannant & Regensburg, 2001, with 10yr design life
- Morgan, Tucker & McInnes, 1994
- Tannant & Regensburg, 2001, with 5yr design life
- Tannant & Regensburg, 2001, with 20yr design life

Chart axes:
- Y-axis: Limiting mictrostrains (ε_{Lim})
- X-axis: Traffic volume (kt/day) x Performance index[1]

Category III
Category II
Category I

Notes
1 Based on acceptable structural performance of road and maximum deflection under fully-laden rear dual, where Performance Index (RPI) varies from
   1 Adequate but fairly maintenance intensive,
   2 Good with normal maintenance interventions,
   3 Outstanding with low maintenance requirements.
2 For Tannant & Regensburg models, design life based on 220 tonne payload truck, load cycles determined using two axles and Performance Index of 2 used.

Haul Road Category	Typical Description	Range of Maximum Recommended Vertical Elastic Strains (microstrains)	
		Traffic Volumes >100kt/day	Traffic Volumes <100kt/day
Category I	Permanent high volume main hauling roads and ramps, in- and ex-pit. Operating life >5 years	900	1500
Category II	Semi-permanent high volume roads in- and ex-pit, dump access roads. Operating life <5 years	1500	2000
Category III	Short-term roads, goodbye ramps and dump finger roads. Operating life <6 mths	2000	2500

Figure 3.22 Typical limiting sub-grade vertical compressive strain values suggested from various sources.

typically be performed once for a mine unless conditions such as truck type and size or material availability would change. For a mine a summary of the mine haul road design for different class roads, shown in Figure 3.25, would form part of the Standard Working Procedures (SWP) file. Such a summary shows the structural layers for every road classification and vehicle combination, as well as construction guidelines.

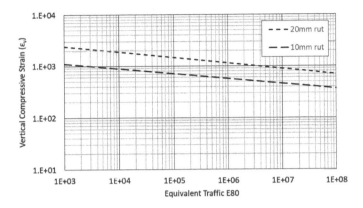

Figure 3.23 Recommended vertical subgrade strain criteria in terms of rut depth.

Source: modified after Jordaan, 1994.

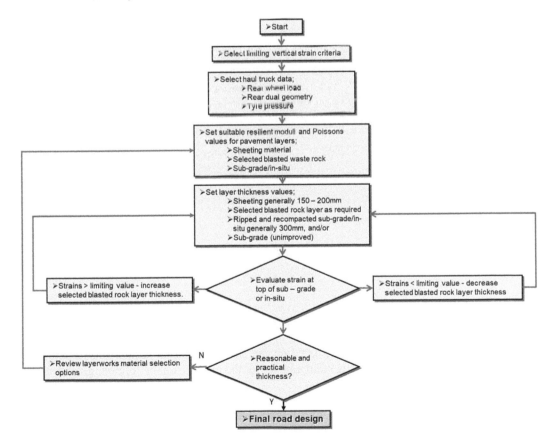

Figure 3.24 Flow diagram for the mechanistic structural analysis.

Table 3.8 Example of a mechanistically derived haul road design for a CAT793D truck, based on three categories of road design and associated limiting vertical compressive strain and vertical deflection constraints.

			CAT793D(laden). Tyres 700kPa. OEM GVM. <100kt/day								
Road design category			CAT I			CAT II			CAT III		
Layer	Description	E_v (MPa)	Thickness (mm)	✿ ε_v (microstrain)	Displacement ⬇ (mm)	Thickness (mm)	✿ ε_v (microstrain)	Displacement ⬇ (mm)	Thickness (mm)	✿ ε_v (microstrain)	Displacement ⬇ (mm)
W/C	Wearing course	350	200		4.4	200		6.5	200		7.3
Base	Selected dump rock	3 000	740			370			290		
Sub-grade	Clay CBR=5 (Infinate depth)	50	inf	908 (Limit 900)	4.1	inf	2001 (Limit 2000)	6.3	inf	2503 (Limit 2500)	7.1

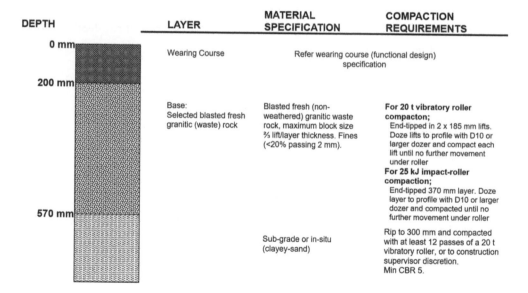

LMN MINE	CATEGORY II HAUL ROAD LAYERWORKS NEW CONSTRUCTION
Traffic volume and operating life	CAT 793D, <100ktpd, medium-term roads (<5yrs)
Sub-grade	In-situ/Sub-grade softs 5 ≤ CBR(wet) ≤ 12

DEPTH	LAYER	MATERIAL SPECIFICATION	COMPACTION REQUIREMENTS
0 mm	Wearing Course	Refer wearing course (functional design) specification	
200 mm	Base: Selected blasted fresh granitic (waste) rock	Blasted fresh (non-weathered) granitic waste rock, maximum block size ⅔ lift/layer thickness. Fines (<20% passing 2 mm).	**For 20 t vibratory roller compacton;** End-tipped in 2 x 185 mm lifts. Doze lifts to profile with D10 or larger dozer and compact each lift until no further movement under roller **For 25 kJ impact-roller compaction;** End-tipped 370 mm layer. Doze layer to profile with D10 or larger dozer and compacted until no further movement under roller
570 mm	Sub-grade or in-situ (clayey-sand)		Rip to 300 mm and compacted with at least 12 passes of a 20 t vibratory roller, or to construction supervisor discretion. Min CBR 5.

Figure 3.25 Example of a structural design guideline summary for use in SWP.

3.3.4 Example applications

Two examples are presented here, based on the structural design required for a haul road operating CAT793D trucks. Data requirements for the models are the following:

• CAT793D operates at OEM GVM, no payload increment, no load transfer and 700 kPa tyre pressure (i.e. 631 kN wheel load and a contact radius of 0.536 m as per Table 3.3).

- An in-situ material of strength CBR 5 and selected blasted waste rock that meets the requirements previously specified. The wearing course material meets the minimum CBR requirements for this layer (CBR > 60). Effective elastic modulus values are 50 MPa, 3000 MPa and 350 MPa respectively, with a Poisson's ratio 0.40, 0.35 and 0.35 respectively for the three layers.
- Layer thicknesses above the semi-infinite in-situ are 370 and 200 mm (with the thickness of the selected blasted rock layer being determined iteratively as per Fig. 3.24).

An example of ELSYM5A input (Table 3.9) and output (Table 3.10) is given and no special formatting of the tables was performed, demonstrating that the 1972 Fortran format was very primitive. (The RH column in each Table has been added to assist in explaining the input and output file formats.)

In Table 3.9, lines 10–17 determine the XY positions of the two laden wheels of the dual, which are located at a centre-to-centre spacing of 1626 mm. Two XY positions are required at which stress, strains and displacements are determined, and as explained earlier is typically under one of the two wheels, or deeper in the pavement, at the centre point between the two wheels (i.e. 1626/2 mm). Lines 18–20 define the Z depths at which the above calculations are performed, in this case at the surface (1 mm) and at the selected blasted dump rock and in-situ interface (571 mm). These Z depths reflect the layer thickness values specified in lines 6–8.

Ensure that the measurement is in the correct layer, as giving the depth as the position of the interface results in the strain in the upper layer being calculated. Normally the interface depth plus 1 mm is used (since internal conversions of units result in minor rounding differences in the output).

Table 3.9 Input information format – ELSYM5A Example.

	line
* ELSYM5 INPUT DATA FILE ELSDAT	1
*	2
* JOB DESCRIPTION	3
Example 1, CAT 793D OEM GVM	4
NUMBER OF LAYERS = 3	5
THICKNESS,POISSEN VALUE,ELASTIC MODULUS 200.00 0.35 350.00	6
THICKNESS,POISSEN VALUE,ELASTIC MODULUS 370.00 0.35 3000.00	7
THICKNESS,POISSEN VALUE,ELASTIC MODULUS 0.00 0.40 50.00	8
LOAD,TYRE PRESSURE,CONTACT RADIUS = 631.00 700.00 0.00	9
NUMBER OF LOADS 2	10
* X POSITION OF LOAD,Y POSITION OF LOAD	11
0.00 0.00	12
1626.00 0.00	13
NUMBER OF XY POSITIONS = 2	14
* XY COORDINATES	15
0.00 0.00	16
813.00 0.00	17
NUMBER OF Z POSITIONS = 2	18
* Z VALUES	19
1.00 571.00	20

Table 3.10 Output information format – ELSYM5A Example.

	line
ELSYM5 3/72–3, ELASTIC LAYERED SYSTEM WITH ONE TO TEN NORMAL IDENTICAL CIRCULAR UNIFORM LOAD(S)	1
ELASTIC SYSTEM 1 – Example 1, CAT 793D OEM GVM	2
ELASTIC POISSONS LAYER	3
MODULUS RATIO THICKNESS	4
1 350. 0.350 200.000 MM	5
2 2999. 0.350 369.999 MM	6
3 50. 0.400 SEMI-INFINITE	7
TWO LOAD(S), EACH LOAD AS FOLLOWS	8
TOTAL LOAD.... 629.94 KN	9
LOAD STRESS.... 699.84 KPA	10
LOAD RADIUS.... 535.30 MM	11
LOCATED AT	12
LOAD X Y	13
1 0.000 0.000	14
2 1625.997 0.000	15
RESULTS REQUESTED FOR SYSTEM LOCATION(S)	16
DEPTH(S)	17
Z= 1.00 571.00	18
X-Y POINT(S)	19
X= 0.00 813.00	20
Y= 0.00 0.00	21
	22
ELSYM5 3/72–3, ELASTIC LAYERED SYSTEM WITH ONE TO TEN NORMAL IDENTICAL CIRCULAR UNIFORM LOAD(S)	23
	24
ELASTIC SYSTEM 1 – Example 1, CAT 793D OEM GVM	25
Z= 1.00 LAYER NO 1	26
X= 0.00 813.00	27
Y= 0.00 0.00	28
NORMAL STRESSES	29
SXX -0.7024E+03 –0.3203E+03	30
SYY -0.7986E+03 –0.4517E+03	31
SZZ -0.6970E+03 –0.1285E+02	32
SHEAR STRESSES	33
SXY 0.0000E+00 0.0000E+00	34
SXZ 0.9194E-01 0.1236E-08	35
SYZ 0.0000E+00 0.0000E+00	36
PRINCIPAL STRESSES	37
PS 1 –0.6970E+03 –0.1285E+02	38
PS 2 –0.7024E+03 –0.3203E+03	39
PS 3 –0.7986E+03 –0.4517E+03	40
THETA 0.2198E+04 0.7848E+03	41
DEV. STRESS 0.1016E+03 0.4388E+03	42
(FOR THETA AND DEV. STRESS COMPRESSIVE IS POSITIVE (+))	43
PRINCIPAL SHEAR STRESSES	44
PSS1 0.5081E+02 0.2194E+03	45
PSS2 0.2691E+01 0.1537E+03	46
PSS3 0.4812E+02 0.6567E+02	47
DISPLACEMENTS	48
UX 0.3776E+00 –0.8329E-08	49
UY 0.0000E+00 0.0000E+00	50
UZ 0.6530E+01 0.6532E+01	51

NORMAL STRAINS   52
EXX -0.5114E-03–0.4509E-03   53
EYY -0.8828E-03–0.9577E-03   54
EZZ -0.4906E-03 0.7356E-03   55
SHEAR STRAINS   56
EXY 0.0000E+00 0.0000E+00   57
EXZ 0.7096E-06 0.9538E-14   58
EYZ 0.0000E+00 0.0000E+00   59
PRINCIPAL STRAINS   60
PE 1–0.4906E-03 0.7356E-03   61
PE 2–0.5114E-03–0.4509E-03   62
PE 3–0.8828E-03–0.9577E-03   63
PRINCIPAL SHEAR STRAINS   64
PSE1 0.3922E-03 0.1693E-02   65
PSE2 0.2077E-04 0.1186E-02   66
PSE3 0.3714E-03 0.5068E-03   67
                                                                       68
ELSYM5 3/72–3, ELASTIC LAYERED SYSTEM WITH ONE TO TEN NORMAL   69
   IDENTICAL CIRCULAR UNIFORM LOAD(S)   70
ELASTIC SYSTEM 1 – Example 1, CAT 793D OEM GVM   71
Z= 571.00 LAYER NO 3   72
X= 0.00 812.99   73
Y= 0.00 0.00   74
DISPLACEMENTS   75
UX -0.2106E+00–0.2081E-05   76
UY 0.0000E+00 0.0000E+00   77
UZ 0.6341E+01 0.6632E+01   78
NORMAL STRAINS   79
EXX 0.4111E-03 0.8641E-04   80
EYY 0.6399E-03 0.6313E-03   81
EZZ -0.2001E-02–0.1694E-02

In Table 3.10 lines 1–11 echo the input data (note the slight rounding errors in the data values). The lines following reflect the output from the elastic multi-layer solution:

- Lines 26–28 reflect the Z-depth (1.0) and the two XY locations at which calculations are made, 0.0 and 0.813 (the latter being the mid-point between the two-wheel loads).
- Lines 30–32 reflect the normal stresses determined at the above locations.
- Lines 33–67 similarly show shear, principal, deviator and principal shear stresses, together with displacement, normal, shear, principal, deviator and principal shear strains.
- Similarly, for the remaining Z-depth of 571 mm, lines 68 onwards would represent the same determinations as above (but in this case, only displacement and normal strain data has been reproduced for clarity in the explanation that follows).

In the output Table 3.10, the information of importance is highlighted, which is the surface deflection (given as displacements UZ), as this provides an indication of the potential deflection bowl, which may impact rolling resistance. Low deflections are aimed at, and high deflections often warrant pavement redesign. For the lower layers the vertical compressive strain (EZZ) at the top of the layer is considered, as this is the critical design parameter. These values are those that have been reflected in Figure 3.25 and Table 3.8 (for Category 2 road design layer thicknesses).

A similar exercise has also been run in Mincad's CIRCLY6 program, which has a purpose built routine to assist with haul road structural design, together with a database of commonly utilised mine haul RDTs. Using the same data as with the ELSYM5A example, together with the CIRCLY 'auto design thickness' option (which automatically calculates the thickness of the selected blasted dump rock layer, based on the specified limiting strain), similar results are seen as shown in Figure 3.26. In CIRCLY, the degree to which the limiting strain criteria

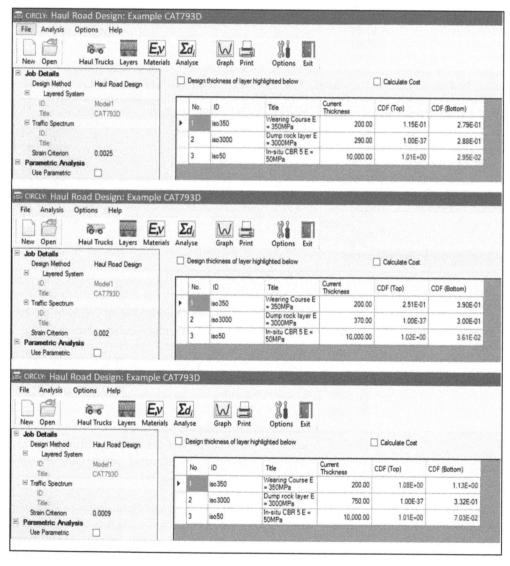

Figure 3.26 Example of output from CIRCLY6 models of haul road design for Category I, II and III designs as per Table 3.8.

is met by the layer thicknesses adopted is referred to as the layer 'CDF' – or cumulative damage factor (defined as the ratio of calculated to the design limiting vertical compressive strains). Thus a damage factor of 1.00 applied to the top of the in-situ layer implies the calculated vertical strain equals the design limiting strain value and thus the design criterion is satisfied.

3.3.5 Comparison of design methodologies

Both the CBR and the mechanistic methodologies have been illustrated by examples. In comparing the two approaches, several key points arise:

*   Referring to Table 3.11 (which reflects the CBR model shown in Table 3.2), there is a significantly larger volumetric and layerworks requirement than for the mechanistically derived equivalent which incorporates the selected blasted dump rock layer immediately below the wearing course. Layerworks thickness is effectively reduced from 1934 to 370 mm (excluding wearing courses). The overall savings in material volumetric and compaction requirements is 1564 mm. When combined with a haul road construction

Table 3.11 Comparison of CBR and mechanistically derived structural designs – layerworks materials and thickness.

			CAT793D(laden)		
CBR-based design			**Mechanistic model results**		
Layer	**Description**	**E_v (MPa)**	**Thickness (mm)**	**ε_v (microstrain)**	**Displacement (mm)**
U Base	CBR >80	350	229	221	6.7
L Base	CBR 60	242	303	2118	6.3
U Sub-base	CBR 28	149	323	2560	5.4
L Sub-base	CBR 16	104	1079	2163	4.1
Sub-grade	Clay CBR=5 (Infinate depth)	50		1363	3.4

			CAT793D(laden). Tyres 700kPa. OEM GVM. <100kt/day		
Road design category			**CAT II**		
Layer	**Description**	**E_v (MPa)**	**Thickness (mm)**	**ε_v (microstrain)**	**Displacement (mm)**
W/C	Wearing course	350	200		6.5
Base	Selected dump rock	3 000	370		
Sub-grade	Clay CBR=5 (Infinate depth)	50	inf	2001 (Limit 2000)	6.3

width of 34 m, this would achieve a saving in material placement and compaction of approx. 53,180 m³/km, or 80%, and also provides a better response to the applied wheel loads.

- The mechanistically analysed alternative also delivers a better response to the applied wheel loads in the sense that the 2000 microstrain limit is not exceeded in any layers. The CBR equivalent, although exhibiting lower strains in the in-situ material, exceeds this design limit in the upper and lower sub-base and lower base layers, as illustrated in Figure 3.27. Whilst the CBR-based method could be appropriate for Category III roads, the excessive material volumetric and compaction requirements indicate the method is also inappropriate under these circumstances.

In the foregoing analysis, it is also important to recognise the role of layer compaction in the development of the design and if different layer materials and resilient moduli are used, the comparative design results will also vary. Additionally, the mechanistic design approach is also sensitive to the assumed depth to bottom of sub-grade material. In this analysis, the depth was assumed infinite, but in many cases can be limited to 5 m (Thompson and Visser, 1996), thereafter a stiff layer is assumed. Critically, should this depth reduce (and the thickness of the sub-grade layer reduces), the stresses and strains induced in this layer by the applied wheel load will increase, requiring a change to the layerworks.

Figure 3.28 shows the comparison of a pavement design determined by the CBR thickness design method, and the preferred pavement composition with a selected dump rock layer located just below the wearing course and is based on the research by Thompson (1996). The figure shows a comparison of the pavement determined by the CBR thickness design method, and the preferred pavement composition based on the research. The two pavements

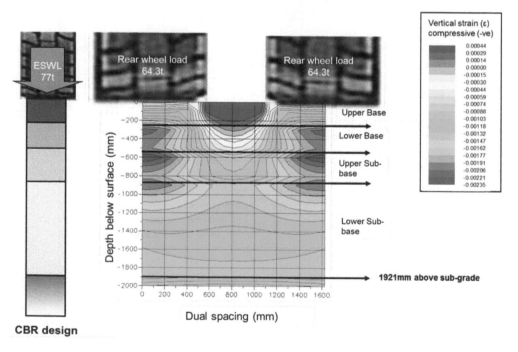

Figure 3.27 Evaluation of CBR derived structural designs using multi-layer elastic analysis.

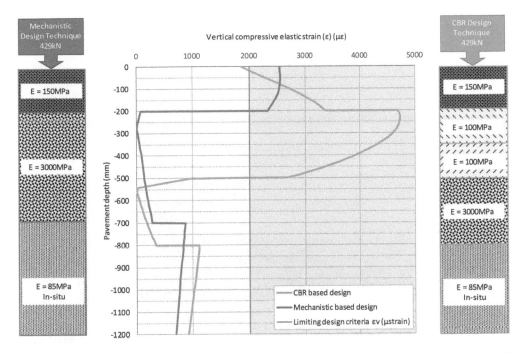

Figure 3.28 Comparison of the vertical compressive strain profile for the mechanistic and CBR thickness design methods.

Source: Thompson and Visser, 1996.

are shown, and the only difference is the base and sub-base layers, the in-situ and the wearing course are the same in each case. The fill layer is a selected blasted waste dump rock with a modulus of 3000 MPa. In the CBR design method, it is placed as fill (300 mm) whereas, with the mechanistic design method, it replaces the base and sub-base of the CBR-method (150 mm each) and has a thickness of 500 mm, which would normally be constructed in three layers (or lifts). The figure shows the vertical compressive strain profile with depth, determined by calculating the strain at depth increments. Note that there are major discontinuities at the layer interfaces, hence the importance of calculating above and below the interface.

Note that the CBR method (blue line), has a vertical compressive strain in the upper pavement layers of 4700 µℇ, which is excessive in comparison with 2000 µℇ, which is used as a general guideline. The dump rock layer has adequate strength and the vertical compressive strain is within limits, and there will be good performance. This finding was discovered during the research phase when a poorly performing road was analysed with the MDD. The upper layers were constructed with granular materials according to the CBR thickness design method. There was significant deflection in these layers. At a depth of about 1 m a layer of dump rock was used as fill, and it was found that below this layer there was hardly any deflection, and thus vertical compressive strains. The stone-on-stone contact was also the philosophy that Macadam, a Scottish engineer, used during the 1850s when streets were started to be surfaced. In Johannesburg, South Africa macadam pavements have carried heavy urban traffic for more than 80 years without structural distress, which supports the original philosophy. Figure 3.29 illustrates the practical application of the design methodology, with a layer

Figure 3.29 Placement of the selected blasted waste rock layer directly above in-situ material, as envisaged with the mechanistic design methodology for mine haul roads.

of selected blasted waste rock placed directly above in-situ as the primary load carrying layer in the design.

3.4 General construction notes – layerworks

Proper planning and design is essential to ensure that the haul road supports the mine plan and mining activities. Planning and geometric design and their impact during the life of a mine were discussed in Chapter 2. Inadequate planning and design would lead to sterilising minerals. Poor geometric design could result in extended cycle times and poor productivity, as well as damage to the vehicle from frequent gear changes and associated road damage. Often an eye-ball alignment results in unsafe conditions. The proper geometric design must be staked out before construction of the earthworks can commence, and in the process any drainage deficiencies can be identified and corrected.

The following notes highlight important points that should be adhered to when the haul road structural design is built. The notes are not exhaustive and should be regarded as supplemental to standard construction procedures.

3.4.1 Preparation of in-situ road bed materials

Once the alignment of the road has been determined and the formation width established, clearing (removal of surface vegetation) and grubbing (removal of subsurface roots) – if necessary – should precede stripping of any topsoil or growth medium. Topsoil should be removed to depths as directed and is generally a scarce resource on mine sites so should be stripped and carefully stockpiled and protected to maintain its viability for later rehabilitation purposes.

Drainage should be established at an early stage to create the new drainage patterns in the vicinity of the road and to divert flows away from the construction area. During construction the road prism should be kept well drained and protected at all times, as unexpected heavy rain on a poorly prepared construction site could lead to severe damage, high cost of restoration and undue delays in the construction programme. All windrows should be opened after construction and in the right direction according to the longitudinal grade that facilitates water flow to be diverted from the lateral drains (Fig. 3.30), so as to prevent concentrated flow on completed layers, but where necessary flat berms should be constructed to prevent undue erosion of subgrade slopes. All permanent drains should be constructed as soon as possible, plus sufficient additional temporary drains as may be necessary to protect the road prism.

The in-situ materials on which the road is placed should be deep-ripped with a dozer to break down any shallow layering. Isolated local concentrations of weak material for which the pavement has not been designed, should be removed to such widths and depths as necessary. Furthermore, an experienced construction foreman may identify unexpected substandard materials which are of a quality that would be detrimental to the performance of the completed road and should be removed. The excavated volume should then be backfilled with selected blasted waste rock material and compacted.

The road bed should be prepared by shaping where necessary and compacting with a 20 t or similar heavy vibratory roller or 25 kJ three-sided impact roller. Water may be added

Figure 3.30 Windrows opened in a road to divert the water flow from road drains to the peripheral drains.

where required to bring the in-situ to optimum moisture content (generally between 8% and 11% for granular materials, but more for fine-grained materials) to assist in achieving the required degree of compaction. Compaction should continue until negligible movement is seen under the roller or the minimum compacted layer thickness and CBR specified in the structural design is achieved. Figure 3.31 illustrates a DCP testing to determine the road bed CBR following compaction. Dry-back will often allow an increase in strength of this layer and as such the construction should be left to dry for 48–72 hours before layers are added above. Ruts and potholes that develop in the road-bed after completion should be repaired and damaged sections of the road-bed reshaped and re-compacted.

The design CBR should be achieved to at least 300 mm depth below top of in-situ/subgrade. For QAQC (Quality Assurance and Quality Control) purposes, the frequency of tests will depend on site supervision, quality and variability of subgrade along road centreline, but typically for in-situ/subgrade materials, tests at 200 m intervals along the centre, left and right carriageway outer wheel paths would be typical.

Construction practice in tropical climates and weak subgrade conditions has been to avoid, wherever possible, side-cut engineered filling. Where engineered filling on a side slope is unavoidable, poor and weak material should as far as possible be removed. In-situ should be cut as individual benches of 3 m width to accommodate standard construction

Figure 3.31 DCP testing along the centre line of a new road to determine CBR achieved in the road bed.

plant, graded slightly to the outslope to prevent ponding and stepped down across the formation width. This will improve layerworks stability to provide a foundation for constructing with engineered filling.

The width of any excavation should allow for a 45° (also known as 1:1) load spread from the toe of any road embankment at ground level to the base of the excavation to allow for side slope instability and to provide lateral support to the placed materials. Where benches daylight to natural ground surface, the bench edges will be unstable, particularly if spoil or fill has intentionally or unintentionally been side-cast and/or bench drainage does not prevent run-off at low points. Batters in the soils encountered across a site in tropical climates will not be easy to maintain in the high rainfall environment. Materials, particularly the low to non-plastic regolith, will be readily eroded by overland water flow with the development of deep steep sided gullies if run-off is not controlled. If saturated, such material is likely to fail.

During compaction, care needs to be taken to prevent liquefaction by ensuring adequate drainage. Water may be added where required to bring the in-situ material to optimum moisture content (generally however, field moisture contents will be higher than optimum and compaction could be difficult before the material drains). Dry-back is unlikely to be an option in a high rainfall regime.

3.4.2 Placing and compaction of material above road-bed

The designs discussed in this chapter make use of a selected (blasted) waste rock layer as the base component of the road. When a selected (blasted) waste rock layer is located immediately under the wearing course, road performance is significantly improved, primarily due to the load carrying capacity of the waste rock layer, which reduces the susceptibility of the softer sub-grade or in-situ to the effects of high axle loads. The rock layer is the primary structural layer and should be placed as near the surface as possible. It serves no purpose to bury it and then cover it with typical gravel and crushed stone layers as is used for public roads. The wearing course is a protective layer that ensures a vehicle-friendly ride and good functional performance. Wear of the wearing course could result in rocks from the dump rock layer being in contact with the tyres, which could result in early tyre failure. There is thus a minimum thickness of wearing course that can be constructed. A minimum practical thickness is 100 mm, which is used on short-term roads. On the more permanent roads a wearing course thickness of 200 mm is placed to avoid having to regravel the road every year, as such a thickness should last at least three years with careful light grading.

On roads that have a good in-situ support such as ramp roads or in-pit roads, ripping and recompacting the in-situ and a dump rock layer may not be necessary if the in-situ materials are capable of carrying the applied loads. Often, a dozer can even out the floor and create the drainage, and a 100 mm wearing course to protect the tyres would be sufficient for these situations.

A large tracked dozer (D9 or larger, 45 t, 300 kW) is used primarily for ripping and shaping sub-grade/in-situ and (if used in the design), the selected blasted waste rock base layers. Ideally the dozer should be equipped to use a GPS and computer-aided earthmoving system or similar to push the material in the road base or in-situ to the required profile which meets alignment design both in the horizontal and vertical planes. A wheel dozer could also be used to assist the track dozer, but not as primary equipment. This is because the material breakdown caused by the dozer tracks is useful in preparing a finish to the layers (especially for in-pit blasted material), an effect not easily replicated by a wheel dozer. The dozer must be able to shape the material on which the road is built. To do this, it must be able to rip the material loose if required, push it to profile (or grade) and remove oversized rocks. It must also be able to open and spread material tipped by dump trucks as part of the road building process. In doing this, the dozer will also start the process of compaction and will form a smooth surface on which the vibratory or impact roller will operate. The larger the dozer, the better the initial strength of the layer will be and compaction requirements will be reduced (but not eliminated).

The dump rock layer is the primary structural element, and consists of competent rock that will not crush or break under the heavy vibratory or impact roller, nor degrade in the presence of water, with a minimum of fine material (preferably less than 20% passing the 2 mm sieve). Fine material may contain clay, and in the presence of water would become a lubricant which will prevent stable stone-to-stone contact. Fine material would also inhibit drainage of the layer. On some projects where weathered rock was present, material smaller than 25 mm was sieved out by means of screens or a grizzly. In the absence of a suitable rock material, the structural layer may be constructed as a cement stabilised crushed stone or gravel layer. Such a layer would be constructed in a maximum of 200 mm lifts. Cement stabilised materials do not have the durability of competent rock and would degrade over time. It may also be more expensive to construct than dump rock.

The maximum size of rock particles is two-thirds of the layer thickness. Larger rocks would result in compaction equipment riding on the rocks and preventing adequate compaction, which translates into settlement over time. Usually the loader is able to separate oversized material from useful material. If there is a large quantity of oversize material it may be prudent to set up a grizzly (a temporary structure with steel bars spaced such that allow only the required size to pass). If occasional large rocks are delivered to the road, then the dozer can separate and push them to the edge of the road. The dump rock is end-tipped and spread to grade and profile, after which (taking into account thickness reduction because of compaction) the layer is ready for compaction.

Compaction is critical to the success of a road building project. With small, light haul trucks, compaction is sometimes not required because the dozer can compact the layers sufficiently. When larger trucks are used, dozer compaction is not sufficient and a large steel drum vibrating roller, impact or grid roller is needed to shake the rock layer down, interlock the particles and compact the layer. In all cases, compaction by haulage and construction equipment alone in insufficient to meet requirements and will result in a poorly performing structure. Figure 3.32 illustrates the difference in wearing course finish achieved with (LHS) compaction from truck traffic only, and (RHS) compaction with steel drum vibratory roller. The road was lightly bladed to reveal the conditions just below the surface and the large areas of loose, uncompacted material are clearly evident, which will lead to rapid deterioration and poor wet-weather trafficability. On a well-compacted road, the aggregate and finer fill material is all tightly bound (so much so that blading actually cuts through the aggregate, not plucking it out of the road). This would be typical of a well-compacted and resilient wearing course.

Hauling, spreading and compacting equipment should be routed uniformly over the full width of the layer to be compacted. Construction specifications for a dump rock layer should state that the layer is end-tipped, dozer shaped to road prism in lifts as directed in the specifications (not exceeding about 200 mm for 230 kN vibratory roller compaction and 1000 mm for a minimum 25 kJ impact roller), and then rolled until negligible movement is seen under the roller. Alternatively, intelligent compaction systems may be used to identify when layer compaction is complete. For an impact roller (Fig. 3.33), generally speaking, 10–15 passes should be sufficient per lift. The layer can be blinded with a 75 mm crusher run of the same

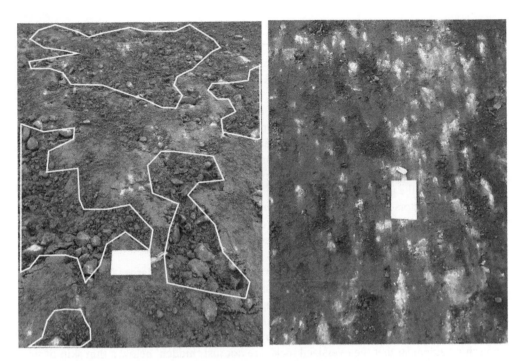

Figure 3.32 Examples of identical haul road wearing course with compaction by trucks only (LHS – soft-spots outlined) and steel drum vibratory roller (RHS).

Figure 3.33 Typical impact rollers used for mine road construction.

Source: images courtesy of Broons.

waste material if required to finish, or a grid roller can also be applied to produce this surface. During the spreading and compaction process the material should be compacted and broken down by means of a vibratory or impact roller in order to achieve a good mechanical inter-lock of the rock and a maximum compaction of the finer material in the interstices between the rocks. No water is required for compaction of this material as coarse material with mini-mal fines does not require moisture for compaction.

It is recommended that construction techniques be developed based on method specifi-cation, using at least a 300 m section of trial road in which at least 90% of consecutive tests satisfy or exceed the performance standard specified.

Most contractors can supply impact rollers – however, it is also a useful piece of equip-ment for a mine to own and operate since it can be used to great effect in preparing waste dump roads, compacting the tip head, and blinding the bench floor in the loading area, which is often an area of potential tyre damage.

Grid rollers should not be used in a primary compaction role. A large vibratory grid roller helps break down larger material. The grid roller is also useful in wearing course prepara-tion, if hard and slightly oversize blocky aggregates are used. The roller will breakdown the blocky material and compact it, resulting in a strong wear and erosion–resistant surface. However, this 'breakdown' does not occur very deep into the layer – so care must be taken if using this equipment that the oversized rocks are not just 'hidden' below a thin skin of finer material. If this is the case, the oversized rocks will soon 'grow' out to the surface and make road blading difficult (due in reality to gravel loss to the roadside during trafficking and con-sequent exposure of the blocky material).

A large grader is used during layerworks construction to open and spread layerworks material prior to compaction, re-shape layerworks following compaction and open or spread crushed rock material as a pioneer or thin 'blind' layer on top of the selected blasted waste. As with the dozer, a laser- or GPS-guided levelling system to assist the operator in keeping grade and profile alignments is valuable (particularly as will be seen in Chapter 4, for wearing course construction and finishing).

Figure 3.34 Grader equipped with GPS-based control system.

Source: image courtesy of Synergy Positioning Systems Pty Ltd.

3.5 Large tracked and platform-type equipment

Large tracked and platform-type equipment include draglines for which walkways may be required (as shown in Fig. 3.35), and truckless (continuous) mining equipment such as loaders and in-pit crushers. Draglines may have a total mass of 3500–6000 t and apply a pressure of 230–270 kPa when moving on its shoes (approximately 80% of total mass), and truckless mining equipment may have a total mass of 2200 ton and apply a pressure of up to 370 kPa. These applied pressures may be compared with the 700–900 kPa tyre pressure of haul trucks. Figure 3.36 shows the configuration of an example of a mobile sizing rig (MSR).

The tracks of this type of equipment are rectangular, and to allow analysis by linear elastic layer packages the rectangle must be represented by equivalent circles that give the same contact area, and thus the same contact pressure, as shown in Figure 3.37.

In the case of trucks, only the dual wheel assembly is analysed in the mechanistic analysis. However, for the crawlers the applied loads have an influence with depth as shown by the superposition in Figure 3.15, and both crawlers need to be input. For ELSYM5a the limitation is 10 circular loads.

The limiting criteria are permanent deformation, and a critical value is 20 mm rutting, or given by a vertical compressive strain of 2000 $\mu\varepsilon$. Note that the design has to be a low-risk one, as a mine cannot afford to be unproductive if the equipment is bogged in the weak materials and cannot work. Although the criteria may be considered to be conservative it should be borne in mind that the drainage of the loading area may be a challenge, as often surface water drains

Figure 3.35 Walkway built for large walking dragline.

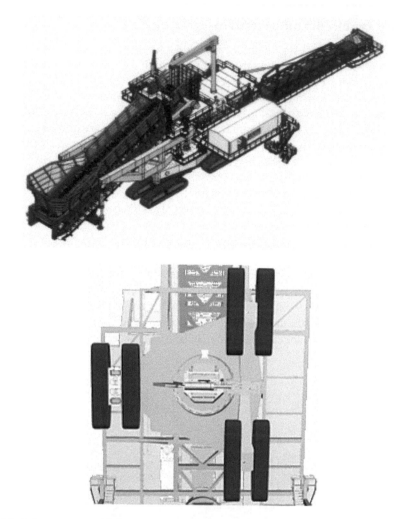

Figure 3.36 Schematic design of the MSR equipment (top) and plan schematic view of the undercarriage and MSR equipment contact area (bottom).

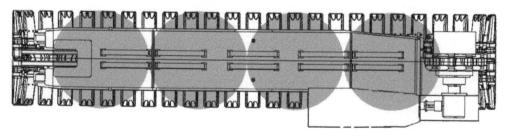

Figure 3.37 Modelling a crawler by using circular loads.

towards the loading face. Under these conditions crawlers will churn up the material and the reworked material would have a lower density and strength than anticipated. Furthermore the MSR is a crusher that also applies vibrations to the soil, whereas the analysis method is mainly a static one. The guideline values may have to be adjusted based on the experience at a particular site with the specific equipment.

Chapter 4

Functional design

4.1 Introduction and requirements for functional acceptability

The functional design relates to providing a surfacing layer, commonly referred to as a wearing course, or sheeting, that has the ideal characteristics at the point where the tyre meets the road. In particular, this relates to a safe, economic and vehicle-friendly surfacing. An ideal wearing course material would fulfil the following requirements:

- Adequate trafficability under wet and dry conditions
- The ability to shed water without excessive erosion
- Resistance to the abrasive action of traffic
- Absence of excessive dust in dry weather
- Freedom from excessive slipperiness in wet weather
- Low cost and ease of maintenance

Poor functional performance is manifest as poor ride quality, excessive dust, low skid resistance, increased tyre wear and damage, and an accompanying loss of productivity due to the rolling resistance increase associated with surface deterioration (or road 'defects'). These and associated impacts on the truck, haul cycle and mine productivity are more fully evaluated in Chapters 5 and 6. Generally, the result of this deterioration is seen as an increase in overall vehicle operating and maintenance costs. Figure 4.1 illustrates two typical examples of functional distress often seen on mine haul roads. Both can be related to poor selection and maintenance of the wearing course material.

The functional design of a haul road is the process of selecting the most appropriate wearing course (or sheeting) material or mix of materials, typically natural gravel or crushed stone and gravel mixtures that are commensurate with safety, operational, environmental and economic considerations. In addition to their low rolling resistance and high coefficient of adhesion, their greatest advantage over other wearing course materials is that roadway surfaces can be constructed rapidly and maintained at relatively low cost. As with structural designs, if local mine material can be used for construction, the costs are all the more favourable. This cost advantage is, however, not apparent in the long term if the characteristics of the wearing course material result in sub-optimal functional performance.

It is imperative that the pavement structural design provides the desired support, as a good wearing course on a substandard structure would result in unsatisfactory performance. The wearing course is not a structural layer and the thickness must be limited, typically

Figure 4.1 Examples of wearing course functional defects: excessive dustiness (LHS) and small potholes leading to poor ride quality (RHS).

200 mm for roads with a life greater than 2 years (Category I and II roads), and 100 mm for short duration roads. Figure 4.2 shows the placement of a selected wearing course material, spot tipped above the prepared and compacted structural layer.

This chapter will deal with the selection and provision of a suitable wearing course material, and the potential benefits of using a palliative in cases where the available materials do not provide the desired performance.

4.2 Functional characterisation

4.2.1 Material properties

It is relatively easy to identify a wearing course that provides excellent performance or deficient functional performance once placed in service, as shown by the examples in Figure 4.3. However, it is desirable to be able to characterise a material before it is placed on the road to avoid costs associated with constructing an inherently deficient wearing course which may include the costs of remediation or even removal. For this reason the wearing course material has to be characterised by conducting tests that determine the size distribution, also termed 'grading analysis', and the Atterberg limits in a soils laboratory. Based on the functional performance related to material properties developed for public roads (CSRA, TRH 20:1990) a similar approach, albeit with adjusted limits, was developed by Thompson (1996) and further described by Thompson and Visser (2000a).

The chart shown in Figure 4.4 shows that the specification is based on a wearing course material shrinkage product (Sp) and grading coefficient (Gc), defined later. Note that Sp

Figure 4.2 Placement of selected wearing course material on a prepared base (structural) layer, prior to opening, spreading, watering, compacting and finally trimming to shape.

relates to the cohesion and plasticity of the material, whereas Gc defines the density and material stability. This specification is based on converted AASHTO metric sieve sizes and if other sieve sizes are used to classify a wearing course material, interpolations will have to be applied to calculate the percentage passing the equivalent AASHTO sieve sizes. The specification is also based on 100% of the material passing (being finer than) the 37.5 mm sieve ($P_{37.5}$) and, as mentioned previously, if $P_{37.5}$ < 100% then the equivalent $P_{37.5}$ at 100% passing will have to be determined (fractions with size less than 37.5 mm are pro-rated as if the $P_{37.5}$ = 100%):

$$Sp = LS \times P_{0.425}$$ Equation 4.1

$$Gc = \frac{(P_{26.5} - P_2) \times P_{4.75}}{100}$$ Equation 4.2

where
LS = Bar linear shrinkage (%)
$P_{0.425}$ = Percent wearing course sample passing 0.425 mm sieve
$P_{26.5}$ = Percent wearing course sample passing 26.5 mm sieve
P_2 = Percent wearing course sample passing 2 mm sieve
$P_{4.75}$ = Percent wearing course sample passing 4.75 mm sieve

In Figure 4.4, the smaller rectangle (green) defines the ideal wearing course properties range, whereas the larger rectangle (yellow) shows the acceptable performance range. The chart shows the anticipated problems when the material characteristics lie beyond the rectangles, and a designer is able to decide whether the anticipated performance can be accepted or another more suitable material needs to be selected. When the wearing course material or mix of materials is sub-optimal, 'functional' defects rapidly form on the road surface and this creates

Figure 4.3 Example of well (top) and poorly (bottom) selected and performing wearing course materials.

safety and road performance problems for the operation, in addition to increased frequency of road maintenance activities.

If the three most critical haul road defects are considered, it appears that mine road-user preference is for much-reduced defects of wet skid resistance, dustiness, and dry skid resistance. This defines the focus point of the specifications to an area bounded by a grading coefficient of 25–32 and a shrinkage product of 95–130, in which the overall and individual defects are minimised (Area 1). Extending this region to provide poorer (but nevertheless operable) performance defines an additional area (Area 2). Area 2 specifications would suit a Category II or III road, from a performance perspective, whilst Category I or II roads would ideally have a wearing course that falls in Area 1.

In addition to the size distribution and plasticity, the maximum size of particles in the wearing course should be 40 mm. This maximum size limitation will reduce the likelihood of tyre damage and also allows maintenance by motor-grader as large particles prevent effective use of the grader, and also initiate potholes as big rocks are pulled out of the matrix.

The specifications in Figure 4.4 should also be evaluated in the light of other material property limits identified as important in functional performance but not directly assessed in

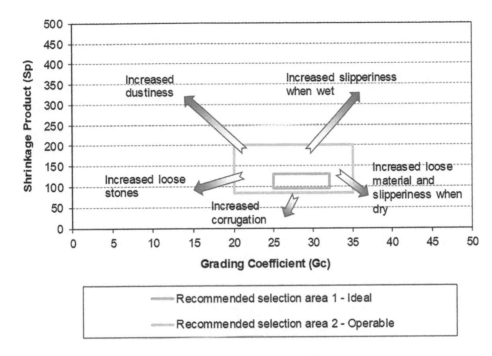

Figure 4.4 Chart for selection of wearing course material.

the selection chart. Table 4.1 presents a summary of these property limits, together with the type of road defects most often associated with departures from the recommended parameter ranges.

Climate is also a consideration in material selection; where a wet climate is encountered, fines ($P_{0.075}$) should be restricted to less than 10% to prevent muddy, slippery conditions when wet. On the other hand, in drier climates, fines should exceed 5% to prevent ravelling or loosening of the wearing course aggregates.

Using the specification in Figure 4.4 and Table 4.1, Figure 4.5 shows examples of wearing course materials (before compaction) representing the following:

- A material which will ravel when trafficked, with a high Gc and low Sp – a 'boney' material deficient in finer size fractions (for binding and cohesion)
- A material which will exhibit dustiness, slipperiness when wet and a propensity to cracking, with a low Gc and high Sp – a 'clayey' material with a large amount of material passing 2 mm
- An ideal material which meets the selection criteria which has all the particle sizes from large to small and not too much fine material (less than 20% smaller than 2.0 mm)

4.2.2 Laboratory characterisation of wearing course

The laboratory tests consist of the grading analysis, Atterberg tests of the liquid limit, plastic limit and linear shrinkage, laboratory compaction test and the soaked California Bearing

Table 4.1 Additional wearing course material selection considerations.

Impact on Functionality Below Recommended Range	Material Parameter	Range Min Max	Impact on Functionality Above Recommended Range
Reduce slipperiness but prone to ravelling and corrugation	Shrinkage Product	85–200	Increased dustiness and poor wet skid resistance
Increased loose stones, corrugations and potential tyre damage	Grading Coefficient	20–35	Increased ravelling and poor dry skid resistance
Reduced dustiness but loose material will ravel	Dust Ratio*	0.4–0.6	Increased dust generation
Increased loose stoniness	Liquid Limit (%)	17–24	Prone to dustiness, reduced ravelling
Increased loose stoniness	Plastic Limit (%)	12–17	Prone to dustiness, reduced ravelling
Increased tendency to ravel, loose stoniness	Plasticity Index	4–8	Prone to dustiness and poor wet skid resistance
Poor wet weather trafficability, churning, excessive deformation and cross-erosion. Maintenance intensive	Soaked CBR at 98% Mod AASHTO	80	Increased resistance to erosion, rutting and improved trafficability
Ease of maintenance, vehicle friendly ride and no tyre damage	Maximum Particle Size (mm)	40	Poor surface finish following maintenance, potholing and potential tyre damage

* Dust ratio defined as $\dfrac{P_{0.075}}{P_{0.425}}$

where $P_{0.075}$ = percentage of material passing the 0.075 mm sieve
$P_{0.425}$ = percentage of material passing the 0.425 mm sieve

Ratio (CBR). The standard values are obtained through the AASHTO/ASTM test procedures, which are used in many countries:

- Particle Size Distribution 75 mm – 75 um AASHTO T87 & 88 / D421 & 422 or local equivalent
- Atterberg Limits PL, LL, PI, and linear shrinkage (LS) AASHTO T 89 & T90 / ASTM D4318 or local equivalent
- Modified Maximum Dry Density (Mod MDD) AASHTO T 180–61 / ASTM D1557 or local equivalent
- CBR 100% Mod MDD AASHTO T 193:/ASTM D 1883 (no soak or 4-day soak depending on local climate) or local equivalent.

Grading analysis

The grading analysis is conducted according to the AASHTO test method T87&88 or ASTM D421&422. The sampling and preparation of the material for testing should be in accordance with national standard procedures or according to the relevant ASTM or AASHTO method. After preparation the material is passed through a set of sieves, as shown in Figure 4.7, with

Gc	Low	High	Within specification
Sp	Low	High	Within specification

Figure 4.5 Potential wearing course materials: on left, poorly graded rounded material (provides poor stability); middle – excessively fine and deficient in larger fractions; right – ideal wearing course with a mix of fine material for binding and larger size fractions for strength and density.

Figure 4.6 Ideal sheeting material following compaction (using Fig. 4.5 RHS example) – note hard road surface.

the largest sieve opening at the top, and the smallest at the bottom, and a pan at the bottom to catch material that passes all the sieves.

Recovering the material retained in each sieve and weighing enables the results to be expressed as the percentage passing by mass, and are typically plotted as shown in Figure 4.8.

Figure 4.7 Mechanical sieve shaker.

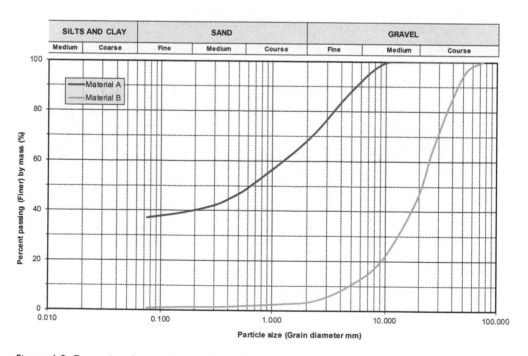

Figure 4.8 Example of a grading analysis of wearing course materials – a fine clayey material A (red line) and a coarser crusher run material B (blue line).

Note that a grading analysis is reported in whole numbers as the precision is not such that decimals can be given. Care needs to be taken with a mechanical shaker shown in Figure 4.7, as overloading of any particular sieve will result in incorrect readings as finer material may be prevented from passing through the sieve. Note that material smaller than 2 mm has to be washed through the sieves in a separate operation as the smaller sieves do not allow all the fine material to pass without washing, and the measured percentage passing is smaller than it should be.

Atterberg limits

The Atterberg limits are the moisture contents in the transition from different phases of a wearing course, as shown in Figure 4.9. These phases relate to material finer than 0.425 mm. Two transition limits are of interest here: the plastic limit (PL) and liquid limit (LL). The moisture content difference between these two states is referred to as the plasticity index (PI).

Plastic limit

As water is added to a dry, hard material it becomes softer, and with the addition of more water becomes plastic, and permits molding. The plastic limit is the moisture content at which the material will crumble at a 3 mm diameter when a thread is rolled. Although Atterberg was an agronomist and developed the empirical test procedure it was found that this limit reflects a shear strength of 170 kPa when determined in a triaxial test (Sharma and Bora, 2003).

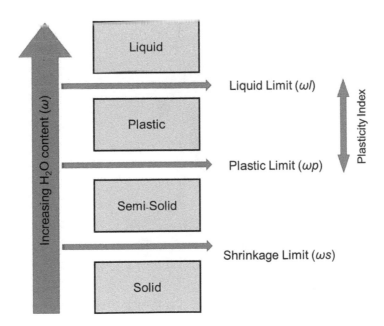

Figure 4.9 Relation of phase consistency and Atterberg limits.

Liquid limit

The liquid limit is the moisture content at which a material passes from a plastic state to a liquid state. The empirical test is conducted with either the Casagrande equipment or the cone penetrometer, shown in Figure 4.10. For the Casagrande equipment, the liquid limit is the moisture content at which the V-groove that is cut into the material in the cup closes over a distance of 10 mm after 25 blows. Since it is difficult to estimate the moisture content at which this happens a three point or single point method is used. The three-point method allows direct reading of the liquid limit, whereas the one point method uses an average

Figure 4.10 Apparatus for liquid limit testing, Casagrande (top) and cone penetrometer (bottom).

Source: images courtesy Humboldt Mfg. Co.

relation between moisture content and its effect on the number of blows to adjust a single measurement. This empirical result is the moisture content which reflects an undrained shear strength of 1.7–2 kPa in a triaxial test (Sampson and Netterberg, 1985; Sharma and Bora, 2003). Note that the base of the ASTM equipment is a hard rubber, whereas the British standards device has a bakelite base which gives a different result.

For the equivalent test using a cone penetrometer, a portion of the material is placed into the test cup and with the tip of the cone just touching the surface of the soil. The cone is released to penetrate into the cup under its own mass in 5 s. The penetration reading can then be converted to liquid limit based on various moisture content factors.

Plasticity index

The plasticity index (PI) which is commonly used in road building is calculated as follows:

$$Plasticity\, Index\,(PI) = LL - PL \qquad\qquad \text{Equation 4.3}$$

Linear shrinkage

At the completion of the liquid limit test the remaining material is placed into a standard metal trough of 10 mm × 10 mm cross-section and 150 mm long. This material is dried in an oven and the linear shrinkage is determined for use in the Sp calculation. In regions where the linear shrinkage test is not performed routinely the linear shrinkage may be estimated as this:

$$Linear\, shrinkage\,(LS)(\%) = \frac{PI}{2} \qquad\qquad \text{Equation 4.4}$$

It should be borne in mind that this estimation is only an approximation, and an actual measurement is preferable.

Patty and ball test

The moisture sensitivity of a material is indicated by the Atterberg limits. However, Nogami and Villibor (1991) developed a simple test for evaluating tropical fine soils which can be used to eliminate potentially unsuitable materials and saves time and cost of laboratory tests.

The test uses two evaluations of minus 0.425 mm material. The one is drying a patty of 40 mm diameter and 10 mm thick which is molded and compacted in a ring. After drying the material is placed on a porous stone or filter paper in 5 mm of water. The capillary action is noted and the hardness of the material over time is tested with a pocket knife or key, as shown in Figure 4.11 (top). Materials which soften or disintegrate are clearly not suitable as a wearing course. The other evaluation is to determine the diametrical resistance of a dried ball of material with a diameter of 20 mm, also shown in Figure 4.11 (bottom). If the material disintegrates when pressed between thumb and forefinger, it does not have dry weather resistance.

Laboratory density

A laboratory density is determined as a benchmark value against which the field density is compared. In addition the moisture content at which compaction should take place is

Figure 4.11 Patty and ball test.

determined. The preferred laboratory compaction test is the modified energy test where a heavy hammer (4.536 kg) dropping through a distance of 457.2 mm is applied 55 times on each of five layers in a 152 mm diameter mould which is 127 mm high (ASTM D1557–70) or AASHTO T180. Because of the mould edge effects, the maximum particles that are used are those passing the 19 mm sieve, which is smaller than the 40 mm maximum size recommended for a wearing course material. There may be a small influence of this change in grading but it is normally ignored.

The theory of compaction will be discussed by using Figure 4.12. For a given compaction energy, when the material is dry the density is low as there is insufficient lubrication to allow the particles to move into a denser packing. As the moisture content is increased, the

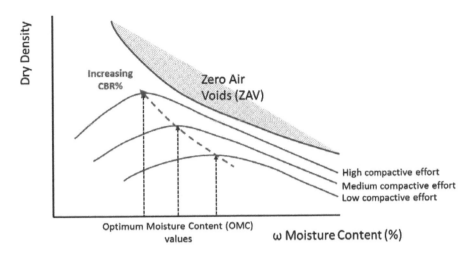

Figure 4.12 Laboratory compaction curve.

particles reorient into a denser packing until the material is at its densest packing at the maximum value of the parabola. This is termed the 'maximum dry density', which is the laboratory standard value, at the optimum moisture content (OMC). Further addition of moisture will push the particles apart and a lower dry density results. The dry density of the wearing course or other constructed layer cannot exceed the zero air voids line (ZAV line), which represents the maximum dry density for a given moisture content. An indication of the support, in terms of the CBR, shows that a moisture content higher than the optimum results in a dramatic reduction in CBR.

It is noticeable that the optimum moisture content reduces with an increase in compaction energy. A higher compaction energy requires less lubrication. Simultaneously the maximum dry density increases.

Usually construction density is specified as achievable at the laboratory optimum moisture content. Experienced construction personnel are able to determine whether the correct moisture content has been achieved by taking a handful of material and squeezing it into a sausage in the hand. If free moisture is evident the material is too wet. If the sausage is tapped and it does not disintegrate the moisture content is correct.

During the wet season it may be difficult to achieve the optimum moisture content (OMC), and usually a construction range of OMC + or – 1% is specified. The resulting lower density must be appreciated over this range. If the material is saturated at the time of opening to traffic severe deformation such as rutting or shoving may occur. When this happens all water spraying must be stopped and the wearing course permitted to dry out. Reworking during the wet season in an attempt to dry out the material may not be successful.

California Bearing Ratio (CBR)

The CBR of a material is the load in kilonewtons, expressed as a percentage of the California standard values, required for a circular piston 1935 mm^2 to penetrate the surface of a compacted material at a rate of 1.27 mm per minute to depths of 2.54, 5.08 and 7.62 mm (ASTM D1883–67). The standard values to these depths are 13.344, 20.016 and 25.354 kN, respectively. The samples are compacted at OMC to the maximum dry density and are then soaked for 4 days to reflect the potentially weakest condition. Note that the removal of particles larger than 19 mm from the test (as discussed in the compaction test) would result that the support value is slightly lower than in practice.

Besides the laboratory test described previously the Dynamic Cone Penetrometer (DCP), introduced in Chapter 3 under structural design, can be used to determine the soaked CBR on the laboratory compacted sample.

4.2.3 Field testing of wearing course

Unlike a dump rock base layer where no compaction control is required other than visual inspection of movement, the wearing course needs to be properly compacted, as a percentage density of the laboratory density, usually 97%, and tested. If natural gravels are used as a base, compaction control, usually 95% would also be required, and tested as for the wearing course.

There are several methods whereby the field density may be measured. The most popular method is the nuclear density device, also known as the hydrodensimeter, shown in Figure 4.13. Generally a 25 mm diameter hole is made by means of a steel bar or drilled to a depth of 150 mm, and the probe of the nuclear device is pushed into the hole. There is a receiver which records the density counts which reduces as the density increases. The moisture content is also measured by recording the hydrogen counts. Since the moisture

Figure 4.13 Nuclear density device used to measure field compaction.

Source: image courtesy Humboldt Mfg. Co.

content is the free water, which is driven off on drying in an oven, there is invariably a difference with the nuclear device moisture content, which also includes water bound in the rock crystals. For this reason a sample is taken from the layer to measure the free moisture content, which is less than the bound plus free moisture content, and will provide a higher dry density value.

Care must be taken when making the hole as a dense layer may be disturbed as evidenced by radial cracks on the surface when the steel bar is hammered into the compacted layer. This means that the layer to be tested is decompacted. For that type of material, a hole must be drilled so that the contractor is not penalised unfairly. Alternatively, the holes can be formed before compaction and reamed as compaction progresses.

There are alternative methods to measure field density such as the sand replacement test (ASTM D1556) and the balloon test (ASTM D2167). These tests involve removing a volume of wet, compacted material 150 mm in diameter and typically 150 mm deep, and measuring the volume with a standard sand or a balloon filled with water. These tests are laborious and slow and are nowadays rarely used, except in cases where there is doubt about the reliability of the nuclear density device.

4.2.4 Combination of materials to achieve desired wearing course properties

Often a single source of wearing course material may not fulfil the requirements for a suitable wearing course material. In such cases two or more materials must be combined to achieve the desired properties. The combination of materials is a linear proportioning of the parameters used to define grading coefficient (Gc) and shrinkage product (Sp):

*   Percentage passing a particular sieve size (Px)
*   Linear shrinkage (LS) or plasticity index (PI)

The calculation of the combined material parameters is achieved from the following equations:

$$P(x)_{mix} = aP(x)_A + bP(x)_B \qquad \text{Equation 4.5}$$

$$PI_{mix} = a(PI)_A + b(PI)_B \qquad \text{Equation 4.6}$$

Where
$P(x)$ is the percentage passing sieve size x
A and B are the different materials
a and b are the proportions of each material such that $a + b = 1$
PI is the plasticity index (%)

The same principle applies if there are more than two materials. Note that if the PI is given as non-plastic (NP) the value in the calculation is zero, and if reported as slightly plastic (SP), then the value is 2. The challenge is to determine the proportions of a, b etc. Generally it is aimed to use two materials, and in exceptional cases three materials, and it should be remembered that it is not possible to use teaspoon quantities, but truckloads. In practice, it is easier to work towards ratios of the two (or three) material types that are easy to implement in the field, such as 1:1, 2:1, 3:1, 4:3 etc.

An Excel spread sheet can be set up where the grading and PI of the two materials are entered into two columns. The values of Gc and Sp are then calculated according to the linear proportioning method for various values of a, remembering that b = (1 – a). By making changes to the combination, again using the proportions of say 1:1, 1:2 etc. as truck loads, the combination that fulfils the Gc and Sp requirements is then found. This method is also applicable to mixing more than two materials.

There is a further procedure which uses linear programming and which can then optimise both the technical requirements of Gc and Sp, as well as cost of the mix if the unit costs of the different materials are known. The technical requirements and cost minimisation are incorporated as objective functions. This cost-optimisation procedure however is not usually applicable in the mining environment, since the primary cost considerations on a mine road are the total costs of building and especially operating and maintaining the road, in which the cost of wearing course material provision is only a small fraction of the total costs.

Table 4.2 shows how two individually unsuitable materials have been combined to form a wearing course material that would meet specification, together with the associated calculations for P(x), PI, LS and Sp and Gc. Figure 4.14 shows the results of the combination (55% Material A and 45% Material B), with the new wearing course material mix meeting the recommended specification for a Category I haul road.

It should be noted that while this approach can be used to approximate the combined material mix specifications, laboratory tests should be conducted on the mix to confirm these estimations – especially the bar linear shrinkage and Atterberg limits estimates, since these are dependent on the medium sand, silt and clay size fraction plasticity which may not conform to the assumed linear combination methodology followed here. In a similar fashion, the CBR (based on the minus 19 mm size fraction) for the combination should also be confirmed in the laboratory.

The grading size distribution chart shown in Figure 4.8 also forms the basis of a graphical procedure (Rothfuchs's method for aggregate blending, CSRA TRH8, 1987) which provides an approximation of the proportions of the materials required in a wearing course. This method is shown in Figure 4.15. It consists of plotting the available materials on a grading analysis chart (or PSD chart – particle size distribution), together with a target grading. Such a target grading is given in Table 4.3 and it has a grading coefficient of 29, which meets the requirements.

The construction is to connect the top and bottom of the two materials where the 2% and 98% lines cut the graph. Note that, for three materials, the values are 3% and 97%. Where the connecting line cuts the target curve a horizontal line is drawn which gives the proportion from the coarsest material to the finest from the bottom. The proportions can then be used to confirm or otherwise that the Sp is achieved.

Table 4.4 confirms that from the horizontal projection of the point at which the 2–98% connecting line crosses the target mix curve, a mix ratio of 58% Material A mixed with 42% Material B would meet the grading coefficient specification, with a value of Gc = 27. Practically, this mix ratio could be adjusted to 50:50 which renders a slightly lower Sp. In practice, traffic- and weathering-induced breakdown of the wearing course often results in an increase in fines and plasticity and this would be reflected in an increase in Sp over time.

It has been shown here how the host rock or source of material used for the wearing course material will significantly influence the grading and the Atterberg Limits. The Atterberg Limits are fundamental to the source material and not readily modified, except through the reduction of the fine fractions. With a wearing course material sourced from crushed

Table 4.2 Calculation of combination of two materials to fulfil the requirements for a suitable wearing course material.

MIX RATIO	55%	45%	FINAL MIX
	Material A	Material B	New Wearing Course
SCREEN ANALYSIS (% Passing) (mm)			
P37.5	100	100	100
P26.5	100	78	90
P4.75	87	12	53
P2	70	5	41
P0.425	45	2	26
ATTERBERG LIMITS			
Liquid Limit (%) (LL)	50	–	20
Plasticity Index (PI)	37	0	20
Linear Shrinkage (%)(LS)	8	0	4.4
Shrinkage Product	360	0	113
Grading Coefficient	26	9	26

$$P(x)_{mix} = a.\,P(x)_A + b.\,P(x)_B$$

$$PI_{mix} = a.\,P(x)_A + b.\,P(x)_B$$

$$SPx = LSx \; \frac{P0.425}{(P37.5/100)}$$

$$GC = \left[\frac{P26.5 - P2}{P37.5}\right] \times \frac{P4.75}{(P37.5/100)}$$

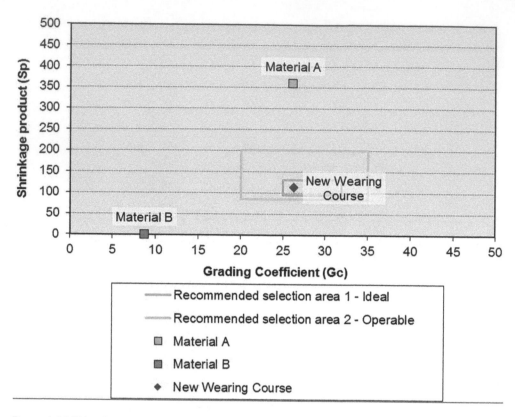

Figure 4.14 Wearing course combinations plotted in comparison to selection criteria.

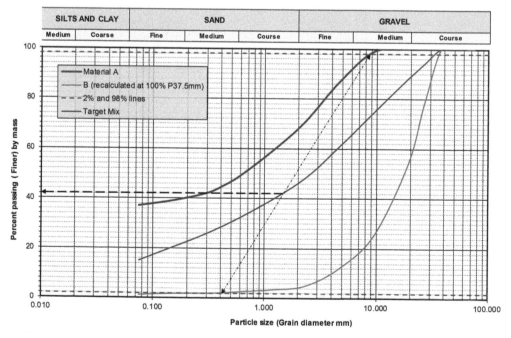

Figure 4.15 Estimation of mix ratios for two materials – graphical construction.

Table 4.3 Target grading for a wearing course which meets the functional requirements.

Sample Name	Material A	Material B[1]	Target Mix
SCREEN ANALYSIS (% Passing) (mm)			
P63mm	100	100	100
P53mm	100	100	100
P37.5mm	100	100	100
P26.5mm	100	78	94
P19mm	100	54	88
P13.2mm	100	29	81
P4.75mm	87	12	61
P2mm	70	5	46
P0.425mm	45	2	29
P0.075mm	37	1	15
ATTERBERG LIMITS			
Liquid Limit (%) (LL)	50	–	
Plastic Limit (%) PL	13	NP	
Plasticity Index (PI)	37	0	
Linear Shrinkage (%)(LS)	8	0	
Shrinkage Product (Sp)	360	0	
Grading Coefficient (Gc)	26	9	29

Note
(1) Material B corrected to 100% P37.5 mm

waste rock, some modification of the product can be achieved to meet a specific sizing or specification by consideration of the crushing process and any subsequent screening applied.

4.3 Benchmarking and monitoring functional performance

In order to monitor and evaluate the functional performance across a network of roads, the functional performance of a road can be tracked by plotting the variation of Gc and Sp over time. This gives some insight into both the current performance of the road from a material specification perspective, but also a functionality and rolling resistance perspective (as will be discussed in Section 4.4). This information assists with making informed road maintenance decisions. In this case, at what point in time should a road be rehabilitated (the wearing course improved to bring it back into specification)? The improvement process basically follows the same process as described in Section 4.2.4, but, in this case, it is required to blend the existing road wearing course with a material which would bring the existing wearing course back to specification.

Figure 4.16 shows a typical analysis of wearing course degradation over time and the associated traffic volumes. Using this information, the rehabilitation needs of the wearing course can be predicted and scheduled based on traffic volumes on the road. In the case illustrated, the wearing course is out of specification (and liable to increased dustiness and potentially slippery when wet) after about five months of traffic at an average of 140 kt/day (over the period Dec to May in the southern hemisphere/wet season). The condition of the wearing course at that point in time (Gc = 198 and Sp = 18) also informs the type of material

Table 4.4 Estimation of combination of two materials to fulfil the requirements for a suitable wearing course material (graphical method).

MIX RATIO	58%	42%	FINAL MIX
	Material A	Material B	New Wearing Course
SCREEN ANALYSIS (% Passing) (mm)			
P37.5	100	100	100
P26.5	100	78	91
P4.75	87	12	56
P2	70	5	43
P0.425	45	2	27
ATTERBERG LIMITS			
Liquid Limit (%) (LL)	50	—	
Plasticity Index (PI)	37	0	21
Linear Shrinkage (%) (LS)	8	0	**4.6**
Shrinkage Product	360	0	**125**
Grading Coefficient	26	9	**27**

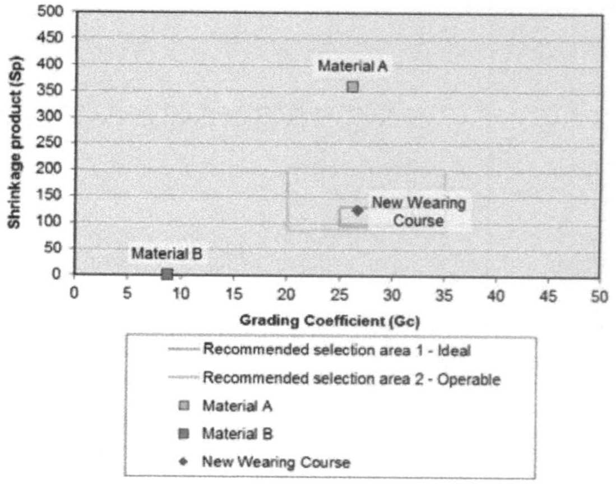

Shrinkage product (Sp) vs Grading Coefficient (Gc)

Material A
Material B
New Wearing Course

Recommended selection area 1 - Ideal
Recommended selection area 2 - Operable
■ Material A
■ Material B
♦ New Wearing Course

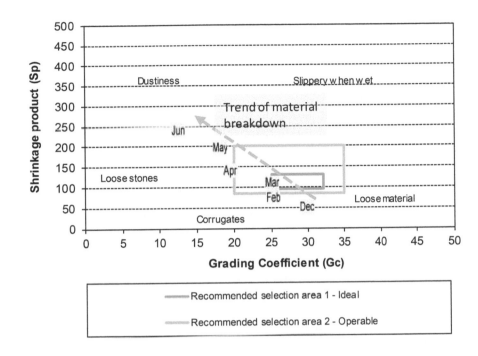

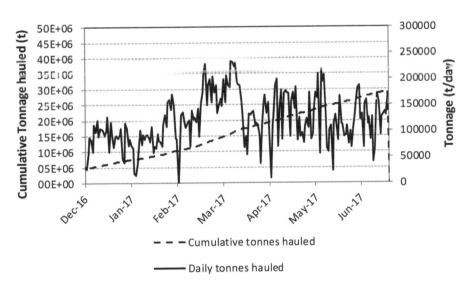

Figure 4.16 Example of wearing course material degradation over time and traffic carried.

sought to rehabilitate the road. In this case, a material with a high Gc (30–35) and low Sp (SP or NP ideally) would be needed.

In the analysis, consideration should also be given to other factors which influence material breakdown, which would vary from road-to-road and between sites. Rainfall especially may accelerate the degradation of the wearing course, as would any fine materials washed or trafficked onto the road from adjacent mining areas. Vehicle speed could also impact the retention of fines (truck wind shear removing fine material from the road surface), as would the frequency of road maintenance – which also contributes both to material mixing and mechanical breakdown.

Whilst the analysis described previously is useful from a wearing course specification perspective, some further insight is useful to evaluate current road conditions and relate the functional defects described previously with changes over time. Two approaches are described here, a qualitative visual assessment methodology and a quantitative modelling approach.

4.3.1 Qualitative assessment of road functionality

Road 'condition' or functionality can be estimated from a qualitative visual evaluation in which the key defects influencing road condition are identified and the product of defect degree (how bad) (measured on a scale of 1–5) and extent (how much) (measured on a scale of 1–5) are scored for each of these defects. The sum of the individual defect scores (degree × extent) gives the overall defect score (DS). The defect score derived for each segment of a haul road network can be used as a basis for the following:

i.   a more quantified and comparable methodology for assessing the maintenance needs of a number of mine haul road network segments, and
ii.   by reviewing records of functional assessments, as an aid to identifying 'under-performing' segments of the network as candidates for rehabilitation.

Thompson and Visser (2000a) analysed the range of defects most commonly associated with mine haul roads, in order of decreasing impact on haulage safety and performance which were typically:

*   Skid resistance – wet
*   Skid resistance – dry
*   Dustiness
*   Loose material (incl. loose stones)
*   Corrugations
*   Potholes
*   Rutting
*   Stoniness – fixed
*   Cracks – slip, longitudinal, and crocodile

To 'rate' these defects in terms of degree or extent, the descriptions in Table 4.5 can be used, or the visual equivalents (only defect degree 1, 3 and 5 given in Figs. 4.17 and 4.18). Note that where the defect is not evident on the road, a defect degree score of 1 and extent score of 1 are used.

Using this approach, a minimum defect score (DS) (representing a road in excellent condition) would be 11 (11 defects × 1 [degree] × 1 [extent]), whilst a maximum defect score (representing a severely distressed road) would be 275 (11 defects × 5 [degree] × 5 [extent]). Individual mines or sites will need to set their own limits of acceptability or maximum

Table 4.5 Description of defect and extent score for functional assessment.

Description of Defect Degree or Extent

Extent Score	1	2	3	4	5
Extent	Isolated occurrence, less than 5% of road affected.	Intermittent occurrence, between 5–15% of road affected.	Regular occurrence, between 16–30% of road affected.	Frequent occurrence, between 31–60% of road affected.	Extensive occurrence, more than 60% of the road affected.
Degree Score	1	2	3	4	5
Degree	**Distress is slight**, difficult to discern and only slight signs visible.	**Distress is easily discernible** but of little immediate consequence.	**Distress is notable** with respect to possible consequence – start of secondary defects.	**Distress is serious** with respect to possible consequence. Secondary defects have developed and/or primary defect is serious.	**Distress is extreme** with respect to possible consequence Secondary defects are notable and/or primary defect is extreme.

allowable DS at which some road maintenance activity is triggered when exceeded. The decision whether or not to maintain the road is not only based on the minimum acceptable DS, but also on selected critical individual defect scores, since these generally adversely affect safety and trafficability.

In tandem with the qualitative evaluation of the defects which control road functionality, it is also important to consider when the road was maintained, since by recording the number of days since routine maintenance was carried out, it is possible to quickly build up a picture of how rapidly the road has deteriorated over time. Also, since functionality and road performance in general is influenced by drainage and erosion, it is useful to assess these two aspects also – poor drainage and/or excessive erosion on the road that would normally trigger some maintenance activity in its own right.

An example of defect degree scoring is given in Figure 4.19. It is important to note that the road illustrated in Figure 4.19 (top) had not seen maintenance for several weeks, compared to the road in Figure 4.19 (bottom) which was recently maintained (bladed) and potholing has re-occurred rapidly. (Rutting occurred and Sp of wearing course too high – hence 'polish' evident after blading.)

Figure 4.20 shows the scoring sheet in which the individual defect degree and extent scores are recorded, the product determined and thus the overall DS found. As mentioned previously, individual mines or sites will need to set their own maintenance intervention thresholds (red, yellow and green bars in Fig. 4.20). This will trigger road maintenance when

DEFECT	VISUAL DESCRIPTION		
	Degree 1	Degree 3	Degree 5
Potholes			
Corrugations			
Rutting			
Loose material			
Stoniness - fixed in road			

Figure 4.17 Functional assessment defect degree visual prompts.

a certain DS is exceeded for a particular road. Typical values from operating experience (mostly operating in dry and temperate environments) are as follows:

- Any single critical functional defect exceeds limits (highlighted yellow)
- Total functional defect score (DS) >130

    - Road maintenance imminent but still trafficable when $60 \leq DS \leq 130$
    - Road in good condition, no immediate maintenance needs when $DS \leq 59$

The decision whether or not to maintain the road is not only based on the total DS, but also on critical individual defect scores too, since these generally adversely affect safety

DEFECT	VISUAL DESCRIPTION		
	Degree 1	Degree 3	Degree 5
Dustiness			
Cracks - longitudinal			
Cracks - slip	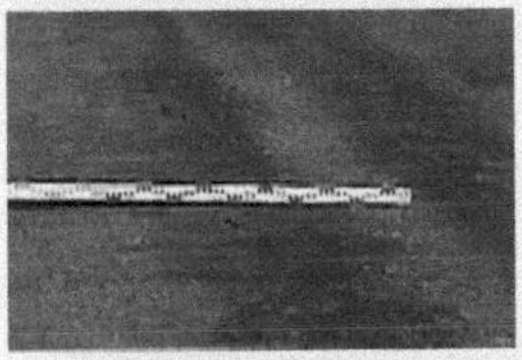	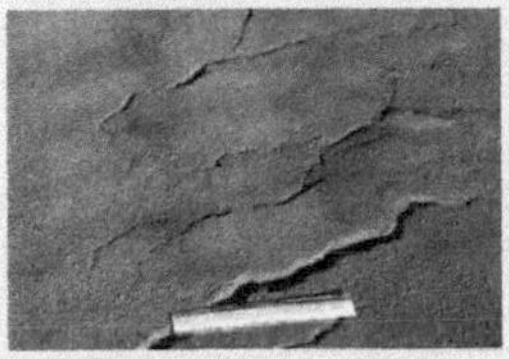	
Cracks - crocodile			
Skid resistance - wet			
Skid resistance - dry			

Figure 4.18 Functional assessment defect degree visual prompts.

TOP: The 'potholes' defect seen in the road would typically score

- a 'degree' of 5 and
- an 'extent' of 2,

giving an individual defect score for pothole defect of (2 × 5) = 10.

BOTTOM: The 'potholes' defect seen in the road would typically score

- a 'degree' of 2 and
- an 'extent' of 5,

giving an individual defect score for pothole defect of (5 × 2) = 10.

Figure 4.19 Functional defect scoring – example application.

and trafficability. These critical individual defects are generally corrugations, loose material, fixed stoniness, dustiness and wet and dry skid resistance and each has a critical functional defect limit which should also be considered in addition to the overall DS. Typical individual DS acceptability limits from operating experience are as follows:

MINE HAUL ROAD FUNCTIONAL AND ROLLING RESISTANCE EVALUATION					
DATE		EVALUATOR			
ROAD		VEHICLE SPEED km/hr (V)			
CHAINAGE		TRAFFIC kt/day			

DEFECT	FUNCTIONALITY			ROUGHNESS (Rolling resistance)		
	DEGREE (1-5)	EXTENT (1-5)	DEFECT SCORE	DEGREE (1-5)	EXTENT (1-5)	DEFECT SCORE
Potholes						
Corrugations						
Rutting						
Loose material						
Stoniness - fixed						
Dustiness						
Cracks - longitudinal						
Cracks - slip						
Cracks - croc						
Skid resistance - wet						
Skid resistance - dry						
TOTAL FUNCTIONALITY DEFECT SCORE (DS) Σ(Defect degree x defect extent)				TOTAL ROUGHNESS SCORE (RDS)		

Road maintenance recommended if: • any **critical** functional defect exceeds limit of acceptability (*) • DS>130	Road maintenance imminent, but road trafficable when total defect score; 60< DS <130	Road in good condition, no immediate maintenance needs when total defect score; DS<60	Refer to graph to read rolling resistance percentages
			ESTIMATED ROLLING RESISTANCE (%)

Comment		
Drainage	On road	
	Side of road	
Erosion	Longitudinal	
	Cross	

Figure 4.20 Functional assessment scoring sheet.

- Corrugations $DS_{max} = 4$
- Loose material $DS_{max} = 4$
- Fixed stoniness $DS_{max} = 5$
- Dustiness $DS_{max} = 4$
- Wet skid resistance $DS_{max} = 9$
- Dry skid resistance $DS_{max} = 9$

However, each mine/site should determine its own critical defect DS limits, based on operating conditions. Figure 4.21 shows an example of a completed functional assessment in which although road maintenance is imminent ($60 < DS < 130$), loose material DS exceeds the 'safe' operating limit of four and thus maintenance would be triggered for this road.

Since functionality and road performance in general is influenced by drainage and erosion, it is useful to also comment on these two aspects – poor drainage and/or excessive erosion on the road would normally trigger some maintenance activity in its own right.

Another approach is to assess road functionality according to Figure 4.22. Functional performance acceptability criteria (limits for desirable, undesirable and unacceptable) should be based on a specific mine's operating experience – average values for many mines are shown in Figure 4.22 – but mostly operating in dry and temperate environments. Used on a daily basis, this chart is useful to record how a road deteriorates over time – a road that always returns values in the red sector is probably a good candidate for rehabilitation. If a road segment is always scoring in the yellow and red sectors – even despite frequent maintenance interventions – then it is worth re-evaluating the wearing course functional design and possibly even the structural and/or geometric designs – since the poor performance is not in itself indicative of poor maintenance – rather an underlying design deficiency. The root cause of a problem needs to be investigated.

This concept can also be useful in day-to-day road maintenance planning. If roads are evaluated at start of shift, they can be marked with red, yellow or green cones to indicate which segments should enjoy maintenance priority (red). This approach is also useful for truck drivers – it helps them anticipate road (and traffic) conditions and thus operate their trucks accordingly. In either case it is important to retain these records and evaluate how each segment of road changes over time and traffic, to identify those segments of the network that continually under-perform, the reasons for this (using the typical defect to identify the root-cause) and thus the most appropriate remediation strategy.

4.3.2 Quantitative assessment of road functionality

Predicting the variation in functionality over time combines key wearing course material parameters with traffic volume and road maintenance interval considerations. Details are presented by Thompson (1996) and Thompson and Visser (2000a). In the models developed particular wearing course materials were selected based on weathering products (as defined by Weinert, 1980) typically encountered in southern African surface mining areas and included pedocretes, argillaceous, arenaceous, basic crystalline and acid crystalline rocks, together with mixtures of these. Climate was discounted as an independent variable since most mines were situated within the same physiographical (Weinert N-value) region of $N = 2$ to $N = 5$.

The development of a predictive model for functionality changes over time is useful for both an understanding of a road's maintenance requirements, and as will be discussed later, as a measure of pavement condition that can be directly associated with vehicle operating costs.

MINE HAUL ROAD FUNCTIONAL AND ROLLING RESISTANCE EVALUATION		
DATE	03/09/17	EVALUATOR	RJT
ROAD	HR1-5WD	VEHICLE SPEED km/hr (V)	14
CHAINAGE	HR1-5WD junction	TRAFFIC kt/day	110

DEFECT	FUNCTIONALITY			ROUGHNESS (Rolling resistance)		
	DEGREE (1-5)	EXTENT (1-5)	DEFECT SCORE	DEGREE (1-5)	EXTENT (1-5)	DEFECT SCORE
Potholes	3	1	3			
Corrugations	1	3	3			
Rutting	2	5	10			
Loose material	2	4	8			
Stoniness - fixed	2	2	4			
Dustiness	2	2	4			
Cracks - longitudinal	2	2	4			
Cracks - slip	1	2	2			
Cracks - croc	2	4	8			
Skid resistance - wet	2	4	8			
Skid resistance - dry	2	4	8			

TOTAL FUNCTIONALITY DEFECT SCORE (DS) Σ (Defect degree x defect extent)	62		TOTAL ROUGHNESS SCORE (RDS)	

Road maintenance recommended if • any **critical** functional defect exceeds limit of acceptability (*) • DS>130	Road maintenance imminent, but road trafficable when total defect score; 60< DS <130	Road in good condition, no immediate maintenance needs when total defect score; DS<60	*Refer to graph to read rolling resistance percentages* ESTIMATED ROLLING RESISTANCE (%)

Comment		
Drainage	On road	
	Side of road	*Drains obstructed by windrows of loose wearing course*
Erosion	Longitudinal	*Erosion in wheel track positions 1, 3&4*
	Cross	

Figure 4.21 Functional assessment worked example.

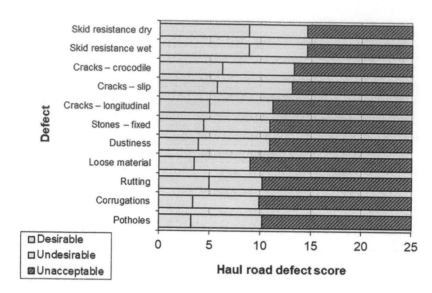

Figure 4.22 Road functionality chart.

Functionality, represented as the defect score (DS) at a particular point in time is a reflection of the type of wearing course material used and its engineering properties, the level of maintenance, season and traffic volumes. Whilst seasonal fluctuation in functionality has been observed, the comparatively frequent watering and blading activities on mine haul roads generally obscure any significant seasonal variations. Thus in the analysis which follows, seasonality is ignored.

A hypothetical defect score progression model is shown in Figure 4.23, based on four distinct traffic, maintenance and wearing course material interactions:

- Immediately following maintenance there will be a traffic induced reduction of loose material and dust defect scores such that the post-maintenance defect scores decrease overall.
- A minimum defect score will be achieved where the progression changes from decreasing to increasing.
- The increasing traffic volumes and dynamic loadings imposed on the road, together with an increase in abrasion result an increase in the defect scores until traffic speed slows and wheel paths change to avoid damaged sections.
- At this point the defect score would remain essentially constant.

A maintenance cycle is defined as starting at point A and finishing between points C and D. The duration, or days between the maintenance cycle varies according to wearing course material performance, typically between 3 and 21 days. Minimum defect scores occur typically from 1 to 3 days after maintenance, depending on wearing course material performance, moisture content and traffic volumes.

The model for functional defect score progression is based on a piecewise combination of two exponential curves, representing the decreasing (LDDD) and increasing (LDDI)

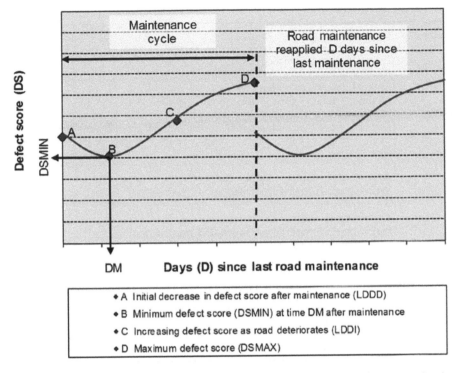

Figure 4.23 Schematic illustration of the development of functional defects on a haul road.

rate of change of defect score with time (or traffic volume). An expression for the minimum defect score after maintenance (DSMIN) and its location in terms of days since maintenance (DM) was also determined. The independent variables defined in Table 4.6 were combined to form the models as shown in Equation 4.5, Equation 4.6 and Equation 4.7. When using these equations, care should be taken to ensure the parameter limits are comparable to the values used in the original research.

$$LDDD = 1.261 + DM(0.000121\ CBR\ KT - 0.0295\ GC + 0.00982\ SP\ DR) \qquad \text{Equation 4.7}$$
$$LDDI = 1.793 + D(0.00228\ KT + GC(0.0103\ DR - 0.01089)) \qquad \text{Equation 4.8}$$
$$DSMAX = 35.025 + 26.783\ M - 0.567\ KT + 1.651\ GC$$
$$+ 0.446\ SP - 10.939\ PI \qquad \text{Equation 4.9}$$

The minimum defect score after maintenance (DSMIN) is given by Equation 4.8 as:

$$DSMIN = 37.915 - 0.158\ KT + 12.709\ M + 1.384\ GC - 0.0875\ SP \qquad \text{Equation 4.10}$$

These models enable the functional response of a mine haul road to be predicted in terms of rates of decrease and, more importantly, rates of increase in defect score with time and traffic volumes. Figure 4.24 compares the prediction model with typical mine site defect

Table 4.6 Definition of parameters and independent variables used in the quantitative modelling of functionality.

Parameter	Description
D	Days since last maintenance
LDDD	Rate of defect score decrease immediately following last maintenance cycle
LDDI	Rate of defect score increase
DSMIN	Minimum defect score in maintenance cycle
DSMAX	Maximum defect score in maintenance cycle
DM	Days between last maintenance and minimum cycle defect score

Variable	Description
KT	Average daily tonnage hauled (kt)
M	Wearing course material type;
	0 = ferricretes
	1 = mixtures and other materials
$P_{0.075}$	Percentage of material passing 0.075 mm sieve
DR	Dust ratio, defined as;
	$$\dfrac{P_{0.075}}{P_{0.425}}$$
PI	Plasticity index
CBR	100% Mod. California Bearing Ratio of wearing course material
GC	Grading coefficient
SP	Shrinkage product,
PL	Plasticity limit

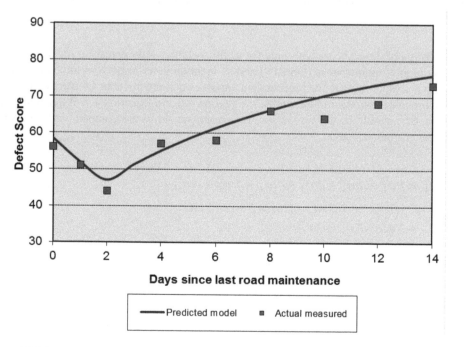

Figure 4.24 Estimation characteristics of functionality model.

score progression data, using Equation 4.5 and Equation 4.6 bounded by Equation 4.7 and Equation 4.8, whilst Figure 4.25 illustrates the effect of traffic volume (kt per day) variation on defect score progression for one particular set of (unimproved) wearing course material properties. As shown, if an intervention level (or maximum acceptable functional defect score, determined by the mine's own operating preference) of 85 is used, a maintenance interval of 5 days is advocated for monthly production of 230 kt or 1200 kt. When the wearing course material is improved, by either replacing or blending to meet the specifications, road maintenance intervals can be extended to every 9 days for 1200 kt per month, and to every 14 days for a road carrying 230 kt/month.

In a similar manner, the impact of changes to traffic volumes on the various segments of a haul road network can be anticipated and managed proactively. However, whilst a model of functional performance is useful, its main disadvantage over the qualitative approach is that it doesn't require the operator to physically assess the road and thus become more familiar with the defects that lead to deterioration – which in themselves are often indicators of how a road wearing course should be improved.

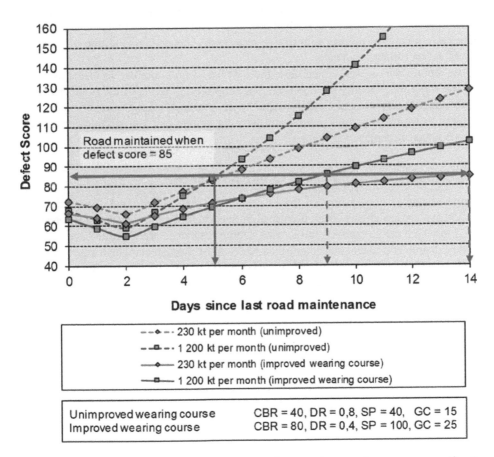

Figure 4.25 Effect of increasing traffic volumes and improving wearing course specifications on functional performance and associated road maintenance intervals.

Another consideration when adopting a functional assessment only as a basis for road maintenance decisions is that although the road would require maintenance from a functional perspective, application of maintenance assets may not in fact generate any cost-savings and may, under certain circumstances, be a waste of time and effort if the 'value' generated by maintenance is not converted into improved productivity, reduced cost per tonne hauled, etc. This aspect is discussed further in Chapter 5. Therefore, although functionality generates some insight into how well a wearing course performs its function – especially in terms of safe and vehicle friendly ride, the economic aspect is more usefully assessed by evaluations based on the concepts of rolling resistance, as introduced in Chapter 1. Functionality and rolling resistance are both related since the defects that contribute to rolling resistance are, in part, many of the functional defects previously discussed.

In the following section, the application of the qualitative and quantitative approaches described previously are now applied to rolling resistance determination.

4.4 Benchmarking and monitoring rolling resistance

In order to make informed road maintenance decisions, some basis of comparison should be established with which to compare segments of road across the network. This comparison is based on the functional defects described previously and – as stated earlier – it is possible to equate *some* functional defects with rolling resistance – hence the condition of a road has a direct effect on rolling resistance.

Two approaches are presented here, the first based on a qualitative visual assessment, using the same functionality methodology but simplified in terms of the number of defects used to evaluate current road conditions and the second method, a *predictive* model of road deterioration which uses truck, traffic and wearing course material parameters to evaluate rolling resistance changes with time.

4.4.1 Qualitative assessment of rolling resistance

Rolling resistance can be estimated from a *qualitative* visual evaluation. A road defect classification system has already been described in Section 4.3.1 and the same approach can be applied in which the key defects influencing rolling resistance are identified and the product of defect degree and extent are determined for each of the defects which exert the greatest impact on rolling resistance. The defects considered to have the greatest influence on mine haul road rolling resistance are these:

i.   potholes;
ii.   corrugations;
iii. rutting;
iv.   loose material; and
v.   stoniness – fixed (in wearing course).

These five defects are evaluated in a similar manner to that described for the functional assessment, except, in this case, a roughness defect score (RDS) is determined as opposed to the functional defect score (DS). The visual prompts for the five defects shown in Figure 4.17 are used, or the descriptions given in Table 4.7.

The RDS (minimum score 5, maximum score 125) must also be converted to an equivalent rolling resistance (RR, %) value, using Equations 4.11, 4.12 and 4.13, or the graph

Table 4.7 Rolling resistance assessment defect degree descriptions.

Description of Defect Degree – Rolling Resistance					
Degree Score	1	2	3	4	5
Potholes	Surface is pock marked, holes are <50 mm diameter.	Potholes 50–100 mm diameter.	Potholes 100–400 mm diameter and influence riding quality.	Potholes 400–800 mm diameter, influence riding quality and obviously avoided by most vehicles.	Potholes >800 mm diameter, influence riding quality and require speed reduction or total avoidance.
Corrugation	Slight corrugation, difficult to feel in light vehicle.	Corrugation present and noticeable in light vehicle.	Corrugation very visible and reduce riding quality noticeably.	Corrugation noticeable in haul truck and causing driver to reduce speed.	Corrugation noticeable in haul truck and causing driver to reduce speed significantly.
Rutting	Difficult to discern unaided, <20 mm.	Just discernable with eye, 20–50 mm.	Discernable, 50–80 mm.	Obvious from moving vehicle, >80 mm.	Severe, affects directional stability of vehicle.
Loose material	Very little loose material on road, <5 mm depth.	Small amount of loose material on road to a depth of 5–10 mm.	Loose material present on road to a depth of 10–20 mm.	Significant loose material on road to a depth of 20–40 mm.	Loose material, depth >40 mm.
Stoniness – fixed in wearing course	Some protruding stones, but barely felt or heard when travelling in light vehicle.	Protruding stones felt and heard in light vehicle.	Protruding stones influence riding quality in light vehicle but still acceptable.	Protruding stones occasionally require evasive action of light vehicle.	Protruding stones require evasive action of haul truck.

presented in Figure 4.26. When using these equations, care should be taken to ensure that the parameter limits are comparable to the values used in the original research:

$$RRMIN = \exp^{(-1.8166 + 0.0028\ V)}$$ Equation 4.11

Where
RRMIN = Minimum rolling resistance at (RDS) = 5
V   = Vehicle speed (km/h)

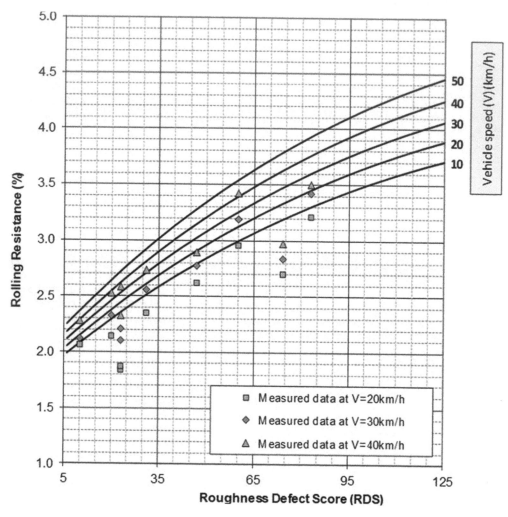

Figure 4.26 Conversion of RDS score to rolling resistance.

$$RRI = -6.068 - 0.00385RDS + 0.0061\ V \qquad\qquad \text{Equation 4.12}$$

Where
RRI = Regression function describing rate of change in rolling resistance

$$RR = 10\left(RRMIN + RDS\ \exp^{(RRI)}\right) \qquad\qquad \text{Equation 4.13}$$

Referring to Figure 4.21 and applying the functional defect scoring for the first five defects, a RDS of 28 results which, at a truck speed of 12 km/h, can be equated to a rolling resistance of 2.4%.

4.4.2 Quantitative assessment of rolling resistance

The rolling resistance (or roughness) of a haul road is primarily related to the wearing course material used, its engineering properties and the traffic speed and volume on the road. These dictate, to a large degree, the rate of increase in rolling resistance. Ideally, road rolling resistance should not increase rapidly – which implies that those road defects (roughness defects) leading to rolling resistance should also be minimised. This can be achieved through careful selection of the wearing course material, which will minimise, but not totally eliminate, rolling resistance increases over time (or traffic volume).

To estimate rolling resistance (RR) at a point in time, an estimate of the roughness defect score (RDS) is required, and this can be determined from an initial estimate of the minimum and maximum roughness defect scores (RDSMIN, RDSMAX), together with the rate of increase (RDSI). Rolling resistance at a point in time (D days after road maintenance) is then estimated from a minimum value (RRMIN) and the associated rate of increase.

Equations 4.14 to 4.17 are used to determine RDS, together with the parameters and variables defined in Table 4.8. When using these equations, care should be taken to ensure the parameter limits are comparable to the values used in the original research.

$$RDS = RDSMIN + \left[\frac{RDSMAX - RDSMIN}{1 + \exp^{(RDSI)}} \right] \qquad \text{Equation 4.14}$$

where:

$$RDSMIN = 31.1919 \ 0.05354.SP - 0.0152 \ CBR \qquad \text{Equation 4.15}$$
$$RDSMAX = 7.6415 + 0.4214 \ KT + 0.3133 \ GC + 0.4952 \ RDSMIN \qquad \text{Equation 4.16}$$
$$RDSI = 1.768 + 0.001 \ D(2.69 \ KT - 72 \ 71 \ PI - 2.59 \ CBR$$
$$- 9.35 \ GC + 1.67 \ SP) \qquad \text{Equation 4.17}$$

Combining Equation 4.14 with Equations 4.11, 4.12 and 4.13 enables an estimate to be made of the progression of rolling resistance as a function of the wearing course material parameters and associated variables described in Table 4.8.

Figure 4.28 shows a typical estimate of rolling resistance (RR determined in N/kg in previous equations, so multiplied by 10 to give rolling resistance as a percentage – RR%) estimated using the equations above and data in the figure.

Figure 4.29 shows the application of the model described here, to determine the typical rolling resistance values experienced across a road or network of roads and the reduction in average rolling resistance achieved through a wearing course rehabilitation project.

4.5 General construction notes – wearing course

In addition to construction loading and hauling equipment and the further requirements for layerworks (presented in Chapter 3), the following additional equipment is typically specified for wearing course construction purposes:

- Water tankers with spray bar for watering during wearing course material compaction. Water must be applied to the loose material being compacted, to bring the material to Optimum Moisture Content (OMC). This is the material moisture content associated

MINE HAUL ROAD FUNCTIONAL AND ROLLING RESISTANCE EVALUATION

DATE	03/09/17	EVALUATOR	EMT
ROAD	HR1-5WD	VEHICLE SPEED km/hr (V)	12
CHAINAGE	HR1 0-670m	TRAFFIC kt/day	110

DEFECT	FUNCTIONALITY			ROUGHNESS (Rolling resistance)		
	DEGREE (1-5)	EXTENT (1-5)	DEFECT SCORE	DEGREE (1-5)	EXTENT (1-5)	DEFECT SCORE
Potholes	3	1	3	3	1	3
Corrugations	1	3	3	1	3	3
Rutting	2	5	10	2	5	10
Loose material	2	4	8	2	4	8
Stoniness - fixed	2	2	4	2	2	4
Dustiness	2	2	4			
Cracks - longitudinal	2	2	4			
Cracks - slip	1	2	2			
Cracks - croc	2	4	8			
Skid resistance - wet	2	4	8			
Skid resistance - dry	2	4	8			
TOTAL FUNCTIONALITY DEFECT SCORE (DS) Σ(Defect degree x defect extent)			**62**	**TOTAL ROUGHNESS SCORE (RDS)**		**28**

Road maintenance recommended if: • any ***critical*** functional defect exceeds limit of acceptability (*) • DS>130	Road maintenance imminent, but road trafficable when total defect score; 60< DS <130	Road in good condition, no immediate maintenance needs when total defect score; DS<60	*Refer to graph to read estimated rolling resistance*
			ESTIMATED ROLLING RESISTANCE (%) 2.4

Comment		
Drainage	On road	
	Side of road	*Drains obstructed by windrows of loose wearing course*
Erosion	Longitudinal	*Erosion in wheel track positions 1, 3&4*
	Cross	

Figure 4.27 Rolling resistance assessment worked example.

Table 4.8 Definition of parameters and independent variables used in the quantitative modelling of rolling resistance.

Parameter	Description
RDSMIN	Minimum roughness defect score immediately following last maintenance cycle
RDSMAX	Maximum roughness defect score
RDSI	Rate of roughness defect score increase
RRMIN	Minimum rolling resistance at (RDS) = 5
RRI	Rate of increase in rolling resistance from RRMIN

Variable	Description
V	Vehicle speed (km/h)
D	Days since last road maintenance
KT	Average daily tonnage hauled (kt)
PI	Plasticity index of wearing course
CBR	California Bearing Ratio of wearing course material (at 100% MDD and 4-day soaked).

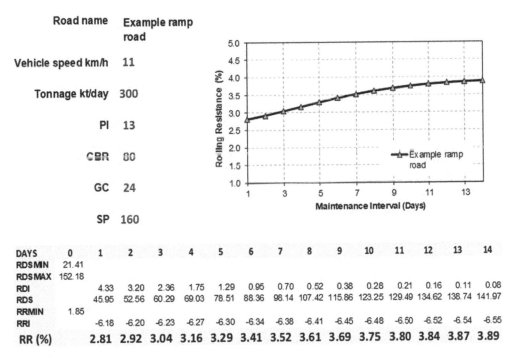

Road name	Example ramp road
Vehicle speed km/h	11
Tonnage kt/day	300
PI	13
CBR	80
GC	24
SP	160

DAYS	0	1	2	3	4	5	6	7	8	9	10	11	12	13	14
RDSMIN	21.41														
RDSMAX	152.18														
RDI		4.33	3.20	2.36	1.75	1.29	0.95	0.70	0.52	0.38	0.28	0.21	0.16	0.11	0.08
RDS		45.95	52.56	60.29	69.03	78.51	88.36	98.14	107.42	115.86	123.25	129.49	134.62	138.74	141.97
RRMIN	1.85														
RRI		-6.18	-6.20	-6.23	-6.27	-6.30	-6.34	-6.38	-6.41	-6.45	-6.48	-6.50	-6.52	-6.54	-6.55
RR (%)		2.81	2.92	3.04	3.16	3.29	3.41	3.52	3.61	3.69	3.75	3.80	3.84	3.87	3.89

Figure 4.28 Estimate of rolling resistance progression using quantitative estimation methodology.

with maximum density and strength. On finished roads, a fit-for-purpose spray-bar is a better solution to effective watering than is a plate or drop spray – coverage is the key issue and no part of the road should be over-watered. Nozzles give finer coverage, less soaking and better watercart efficiency. These dosage and pattern spray systems ideally require a pump with an integrated vehicle speed-delivery control to maintain approx. 0.25–0.5 litres/m² (0.25–0.5 mm film thickness per m²) rates.

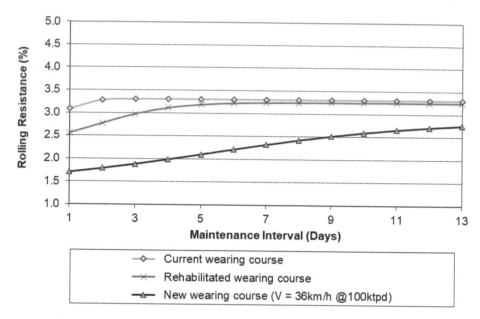

Figure 4.29 Effect of different wearing course options (current, rehabilitated or new mix of materials).

- Mobile crusher for crushing selected waste rock in the event that suitable materials are not available from borrow pits. The use of impact or gyratory crushers reduces the tendency to produce flaky material and should be selected for the crushing of material intended for use as wearing course aggregate. Jaw crushers are the least suitable for materials prone to producing flaky material. Also the crushing process itself, in terms of whether the crusher jaws are fully or partially fed during operation, will influence the final product. In general, for best results, the crusher jaws should always be fully fed during operation.
- An 8–10 m wide offset disc harrow (Fig. 4.30), used for scarifying and mixing wearing course materials. A four-wheel drive tractor tow unit (minimum 25 kW per meter harrow width) is used with the plough. Where a mix of two or more materials is required to make a suitable wearing course material, mixing is very important and the offset disc is the quickest way to achieve this. It can also be used to break down a wearing course layer that has become uneven (after a grader has ripped the road), prior to reshaping with the grader and recompacting. When a grader has ripped the wearing course as part of the rehabilitation or regravelling work, the offset can also be used to break down the wearing course layer, prior to reshaping with the grader and recompacting.

4.5.1 Material selection, crushing, blending and QAQC – wearing course

Placement of the wearing course material (selected and/or blended as discussed earlier) generally proceeds in two lifts (according to design requirements) compacted to

Figure 4.30 Offset disc harrow, used for scarifying and mixing wearing course materials.

95% Maximum Dry Density, as determined by Modified Compaction Energy, to AASHTO T193 (or equivalent) with moisture content at or slightly dry of the optimum moisture content (i.e. -3% to +1% OMC) to give a minimum CBR of 80% using between four and eight passes of a 230 kN vibratory roller.

Wearing course material is spot tipped, opened and mixed. Lower lift of wearing course is placed and shaped to final profile. The first pass of the roller should be on the outside edges of the layer to be compacted progressing towards the centre, overlapping the previous pass by 10%. However when rolling deep loose material all passes in a series should be overlapped by at least half a width of the roller. The gradual extension of the roller onto the unrolled surface makes it possible to apply a concentrated compactive effort on ridges and high spots and keeps the roller running at a truer surface shape. When rolling a layer with a cross-slope, compaction should always commence from the lower edge up towards the centre of the road, because the lower edge of the drum has a tendency to push down the cross-slope. In working towards the crown, wearing course creep (a tendency which can be best resisted by compacted material) will be towards the crown which helps to preserve the shape of the layer. A crowned road is rolled initially starting one edge and working up until the centre is reached. Rolling is then continued by moving diagonally to the opposite edge of the road and working up from there. When the initial rolling pattern has been completed rolling may commence at one side and continue to the other side. When this is done it is imperative that the roller does two passes on the outside edges in each pattern.

In addition to the rolling pattern adopted, correct and uniform moisture content is important for compaction purposes (with/without additional stabiliser or palliative treatment). The recommended practice is to establish wearing course treatment during the trial construction process. However, where excessive and differential settlement of the road is likely, it would be prudent to postpone treatment until settlement is achieved (typically 2–3 months following construction). Water/treatment can be applied once the material is bladed open following which the material can be mixed in with a grader or ideally, offset disc harrow. When too wet for compaction it should be harrowed and allowed to dry before attempting construction. If the layer is too thick it may be necessary to cut half the layer to the side of the road, water the lower half and remix and compact prior to cutting in the remaining half, watering and recompacting.

For QAQC purposes, field testing can be carried out using nuclear densimeters or sand replacement tests (see Chapter 3). Frequency of tests will depend on site supervision, quality and variability of material source (crusher run or borrow-pit) variations:

- For crushed and blended materials, one test per approximately 4000 m^3 placed
- For borrow-pit sourced materials, one test every 2000 m^3 placed

Since method-based specifications are used, once a test section is constructed and the correct method identified from field compaction tests, further road building will only require application of the same method to achieve required results.

4.6 Stabilisation and dust palliation

4.6.1 Products available for stabilisation and palliation

There is often confusion about the desired effect of an additive. A range of products is available to either improve the strength of a material or to serve as a dust palliative. Some products serve both functions. For this reason there must be clarity about the purpose for using an additive. Table 4.9 shows the main purpose of different families of additives.

Table 4.9 Applications of different additives.

Additive	Stabilisation	Dust palliation
Water/ wetting agents		☑
Hygroscopic salts		☑
Natural polymers (lignosulphonate)		☑
Modified waxes		☑
Petroleum resins		☑
Enzymes or biological agents (molasses)	☑	☑
Synthetic polymers	☑	☑
Tar or bitumen products	☑	☑
Sulphonated oils	☑	
Lime or cement	☑	☑

It should be appreciated that the reactions of the different additives depend on the material to be treated. Lime, for example, would have no reaction with sand, but it is highly reactive with clay. Furthermore, the family name is generic, such as a synthetic polymer which is a combination of monomer chains, and the resultant effectivity depends on the polymer composition. For many of these additives there is no standard test procedure, unlike for bitumen where the properties and composition are specified. This means that a client does not have any quality control except that a manufacturer produces according to some ISO guideline, which does not necessarily mean the product composition, which is generally a trade secret. This situation holds potential implications for a client over the long term and alternative control processes such as measuring the strength characteristics of an additive mixed with a standard soil or gravel should be used.

The choice of using an additive is dependent on whether this is a last resort, as in most cases improving the wearing course material would provide an economic solution. Figure 4.31 shows the modelled life cycle costs over a 10-year life, relative to the costs of water-based palliation (for the existing wearing course) at Year 1. The existing situation was a wearing course material that did not meet the functional requirements, and the cost index using water palliation over 10 years escalated to 15. Using a chemical additive reduced the costs index to about 12, which is a significant saving. However, improving the wearing course material to the desired functional requirements resulted in a cost index of 7, which was a further significant saving. Using the same chemical additive on this improved wearing course had a minimal further reduction in the cost index. The first approach is to find a solution that encompasses a wearing course material that fulfils the requirements, and if not suitable, then additives can be considered.

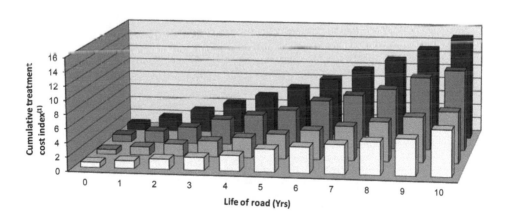

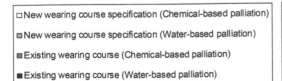

| □ New wearing course specification (Chemical-based palliation) |
| ▨ New wearing course specification (Water-based palliation) |
| ▪ Existing wearing course (Chemical-based palliation) |
| ■ Existing wearing course (Water-based palliation) |

NOTES
(1) Cost index based on unit cost of water-based palliation applied to existing wearing course material.
Deterioration of wearing course over time and traffic not modelled.
Excludes cost of new wearing course reconstruction.

Figure 4.31 Life cycle cost analysis of four scenarios.

4.6.2 The use of water for palliation

Water is recognised as the cheapest treatment for temporary dust reduction (ARRB, 1996). However, in the case of mine haul roads, the frequent re-application rates and capital and operating cost of equipment used, together with (in some cases) the scarcity of water and/or high evaporation rates, may result in water being the least cost-effective option for mines to use for dust suppression.

In many surface mining operations, whilst rolling resistance is often the primary cost-driver in haulage operations, considerable effort and expense is also incurred in the control of dust generation for safety and health reasons. In fact, in countries that have environmental legislation that limits fugitive dust emissions, production often has to be curtailed to meet short-term emission limits.

Particles that become suspended for a noticeable length of time [such as shown in Fig. 4.32(a)] are generally <30 μm in diameter and the proportion of material in this range is approximately proportional to the wearing course material's erodibility. However, this size fraction cannot simply be removed from the wearing course because some fine material is necessary to provide cohesion and bind the larger size fractions of the wearing course, without which ravelling, loose material (including stoniness with the potential for tyre damage) and high rolling resistance would result. Erodibility is reduced by cohesion, which increases

(a)

(b)

Figure 4.32 Watering may not always be effective (a), and overwatering can cause potholes (b).

with clay content and/or the use of additional chemical binders. This forms the basic motivation for the use of some additional agent to reduce a material's inherent erodibility, since the finer fraction, although contributing to cohesiveness, also generates much of the dust, particularly when the material is dry. The presence of larger fractions in the material will help reduce erodibility of the finer fractions, as will the presence of moisture, but only at the interface between the surface and the mechanical eroding action – hence the water-based dust suppression techniques used most commonly on mine haul roads.

However, excessive watering, in addition to increasing operating costs and potentially contributing to increased road-user hazards, can lead also to erosion of the wearing course, and where the material Sp is high, small (3–7 cm diameter) potholes are likely to form, as shown in Figure 4.32(b). This is not problematic per se, but they will induce more rapid wearing course deterioration.

More effective watering can be achieved by using a spray-bar and nozzles mounted close to the road surface, for a more even, lighter watering of the road than would be achieved with a drop-plate arrangement since nozzles give finer coverage, less soaking and better watercart efficiency. Results can be further improved by applying dosage and pattern spray control systems which require a pump with an integrated vehicle speed-delivery control to maintain approximately 0.25–0.5 litres/m² (0.25–0.5 mm film thickness) rates.

Further improvements can be made by adopting different spray 'patterns' and rates of application – this helps reduce potential damage to the road from excessive water (especially on ramps; it also prevents excessively slippery conditions). With a typical 'spot' or 'chequerboard' spray pattern as illustrated in Figure 4.33, moisture from the wet area is carried over to the dry area allowing a larger area to be dust-controlled. However, where traction or friction supply is problematic (poor wearing course material selection), this method of intermittent watering could lead to excessive wheel slip too – especially at the bottom of ramps (where friction supply is at a premium for descending trucks). Light watering improves water car spray productivity and reduces erosion of the road surface.

Figure 4.33 Illustration of the chequer-board pattern of spraying.

More advanced systems, using automatic control, geo-fencing etc. helps reduce over-watering on ramps and an asset management and location system on water-carts is useful to manage spray coverages, optimise vehicle utilisation (spray-time) and as a means of reducing road network dust generation and providing records of palliation activities as may be required as part of a licence to operate. Such a system is illustrated in Figure 4.34, showing geofencing and the resultant road watering and application rate history recorded by the system.

Groundwater, recovered from the mining process or pit sumps is often used to water the haul road and is thought to be slightly more effective than fresh water due to the presence of small amounts of deliquescent salts. However, high concentrations of salts may lead to efflorescence and the salts may only be hygroscopic at relatively high humidities. In addition, some mines report that the roads treated with water containing salts become excessively slippery when wet. However, such reports often fail to specify the wearing course material to which the water was applied, hence the difficulty in identifying the real causes of such slipperiness but in general terms these effects are thought to be associated with excessive clay fractions in the wearing course.

Where watering alone is insufficient to reduce dust to acceptable and safe emission levels, the mechanism of dust formation on a mine road must be considered. Dust generation is the process by which fine wearing course material becomes airborne. Such generation is termed a fugitive (or open) dust source. The amount of dust that will be emitted is a function of two basic factors:

i.   the wind-erodibility of the material involved; and
ii. the erosivity of the actions to which the material is subjected.

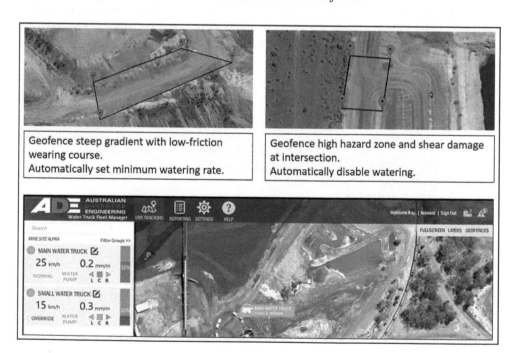

Figure 4.34 Example of automated control of water application rates using geofencing and asset management and location systems' integration.

Source: images courtesy of Australian Diversified Engineering ADE EcoSpray Systems.

In broad terms, the effectiveness of any dust suppression system is dependent on changing material wind-erodibility or erosivity. The wearing course silt and fine sand fractions (i.e. 2–75 μm) are a good indication of its erodibility.

The potential for an activity to generate dust depends on a number of factors, including the following:

- The mechanical actions involved
- The amount of energy imparted to the material
- The scale and duration (frequency) of the activity

Mechanical action involves a combination of reducing particle sizes by impaction and friction, followed by ejection into the air. In the case of mine haul roads, vehicle disturbance can lead to significant wind-related emissions from a surface by

i.   physically ejecting particles from the surface by the action of the wheels; and
ii. creating local turbulent eddies of high velocity.

Thus the amount of dust generated from a pavement surface can depend on the following:

- Wind speed at the road surface. Addo and Sanders (1995) report that speed appears to be linearly related to the amount of dust generated (for light passenger vehicles), as does the vehicle aerodynamic shape, especially the wind shear (lower vehicles with many wheels tending to cause an increase in dust).
- The traffic volume, or number of vehicles using the road.
- Particle size distribution of the wearing course.
- Restraint of fines. This is related to compaction of the road surface, cohesiveness and bonding of the surface material, durability of the material and the amount of imported fines (spillage) on the road.
- Climate, particularly humidity, number of days with rain, mean daily evaporation rates and the prevailing wind speed and direction.

In the surface mining environment of southern Africa for example, many of the previously mentioned factors combine to create a significant dust problem on unpaved mine roads. The extent of the problem has been noted by various authors (Amponsah Dacosta, 1997; Jones, 1996; Simpson et al., 1996; Thompson et al., 1997). In general terms these include the following:

- Loss and degradation of the road pavement material, the finer particles being lost as dust and the coarser aggregates being swept from the surface or generating a dry skid resistance functional defect.
- Decreased safety and increased accident potential for road-users, due to reduced or obscured vision and reduced local air quality.
- Higher vehicle operating costs, with dust penetrating the engine and other components resulting in increased rates of wear and more frequent maintenance.
- Potential loss of production due to shutdowns required to limit emission maximum thresholds.

More specifically, the SIMRAC report by Simpson et al. (1996) investigating the causes of transport and tramming accidents in South African mines highlighted the fact that 74% of the accidents on surface mines were associated with ore transfer by haul truck and service vehicle operation. Dust generation was identified as a significant contributory factor in a number of these incidents. Further work by Thompson (1996) confirmed these findings for vehicles operating on unpaved mine haul roads.

The broader environmental effects of dust have also been reviewed, both from the perspective of unpaved public and mine haul roads. Of particular importance is the finding of Amponsah-Dacosta who conducted an emission inventory for a South African coal strip mining operation. The emission inventory was based on a characterisation of specific dust sources over a specific interval of time, to produce a dispersion model to enable predictions to be made concerning ambient pollution levels and the identification of major control areas. The analysis, conducted according to USEPA (1995) guidelines, found that 93% of the total emissions from the mine were attributable to dust generated from the mine haul road (the second highest, at 2.7%, being attributable to top soil removal). Although a high tonnage operation, the extent of the road network on the mine was similar to other such operations and it was concluded that emissions from the road network would be typical of most opencast coal mines, when calculated on a percentage of total emissions basis.

To reduce fugitive dust emissions from a mine haul road, the broad approach would be to:

• consider modifications to the wearing course material or
• the erodibilty or erosivity of the truck-road surface interaction.

The following actions can also deliver reduced emissions:

• Minimise truck spillage
• Reduce likelyhood of tracking fines onto the road surface, from truck tyres or other sources (poor drainage especially)
• Application of a seal to the road surface
• Use of a tightly bound, high strength wearing course material
• Armouring the surface (placing a thin layer of higher quality wearing course on the existing material or tyning this into the top 50 mm of material)
• Regular light watering of the road
• Use of various chemical dust palliatives
• Reducing vehicle speed
• Modifying engine/retarder blower configuration to blow above, not under the vehicle

Of the approaches listed, only regular watering, the application of chemical dust palliatives together with brooming and the optimal selection of wearing course materials are often the only viable alternatives in controlling mine haul road dust emissions.

4.6.3 Water-spray systems performance modelling

Judicious watering assists in dust suppression, maintaining compaction and therefore strength of the wearing course, in addition to reducing the potential loss of wearing course material. Although watering itself is often seen as a cheap and simple approach to dust suppression, equipment and operating costs often escalate when moisture retention on mine roads is poor, more so during adverse conditions where a combination of high temperatures, high wind speeds and low humidity are prevalent. The degree of dust palliation achieved with watering is a function of the following:

- The amount of water applied per unit area of road surface
- The time between re-applications
- Traffic volumes
- Prevailing meteorological conditions
- The wearing course material
- Extent of water penetration in to the wearing course

Thompson and Visser (2002) described tests conducted to evaluate the efficiency of a water-spray-based dust suppression system. Initially, as a result of the large number of variables affecting the generation of dust, a visual classification system was developed for the 'degree' of dust defect based on the road user's experience from the point of view of a haul truck travelling at 40 km/h. Table 4.10 gives these descriptions, both as they relate to measurement (dust reading in mg/m³ of the minus 10 µm dust fraction) and practically to what the mine would visually experience. In general, the consensus was that a dust defect score of 2 would represent a typical dust defect intervention level (where some dust suppression activity would be initiated). This defect score was based primarily upon the visual effects (road safety and driver discomfort), rather than any perceived health impacts.

By recording the dust plume generated by a haul truck as it passed at a set distance and speed from the monitor, analysis of the data enabled a first estimate to be made of the time taken for the degree of palliation to decay to zero and the effect of evaporation rates on this time. Figure 4.35 illustrates a typical dust-time curve from a particular test site, showing how dustiness increases with each vehicle pass from dust defect degree one (immediately following spraying) to degree four 90 minutes after application.

An initial estimate of the time to no effect of palliation (from when the road was water-sprayed at recommended application rates) was found, assuming (initially) that dust generation is independent of vehicle shape and aerodynamics (these effects being analysed in isolation later). Regression of time to no effect of palliation on monthly evaporation rates for stations in Weinert's climatic region N = 2–5 (Weinert's N-value describes the durability of natural wearing course material, based on the relationship between calculated evaporation rates (for the warmest months of the year) and the average monthly rainfall). Equation 4.16 estimates this time.

$$X_0 = 286.8 - 0.73 E_m \hspace{4cm} \text{Equation 4.18}$$

Where
X_0 = Time to no effect of palliation (mins)
E_m = Average monthly evaporation rate (mm/month) for climatic region N = 2–5

In the climatic region described previously, if time to no effect of palliation is considered in terms of the percentage of total dustiness, in typical winter conditions, for an average of 50% palliation, re-application is required at approximately 3-hourly intervals whilst in summer, the frequency is increased to approximately 1½-hourly intervals. These

Table 4.10 Classification of the degree of haul road dust defect.

Typical dust defect photograph showing haul truck travelling past monitor point at 40 km/h	Dust defect degree and associated peak dust levels (approx. mg/m³ for −10 µm dust per haul truck pass)	Qualitative dust defect degree descriptions
	Degree 1 <3.50	Minimal dustiness
	Degree 2 3.51–23.50	Dust just visible behind vehicle
	Degree 3 23.51–45.00	Dust visible, no oncoming vehicle driver discomfort, good visibility.
	Degree 4 45.01–57.50	Notable amount of dust, windows closed in oncoming vehicle, visibility just acceptable, overtaking hazardous.
	Degree 5 >57.51	Significant amount of dust, window closed in oncoming vehicle, visibility poor and hazardous, overtaking not possible.

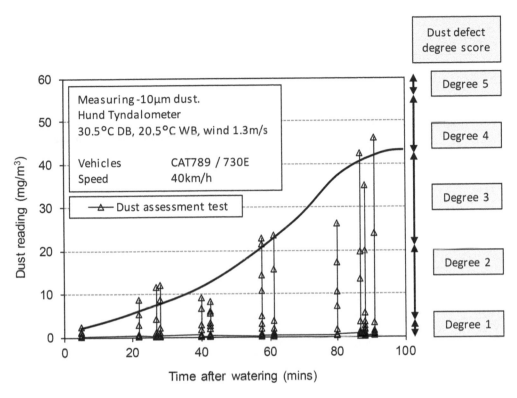

Figure 4.35 Rate of increase in dustiness following watering on a mine haul road.

rates are based on an average of 50% palliation which does not accommodate the road-user preferred dust defect limit of degree 2, corresponding to a maximum dust concentration of 23.5 mg/m³.

To provide an initial estimate of the dustiness associated with a particular wearing course material, Thompson and Visser modelled three parameters;

- Mass of dust as loose material on the road (g/m²) (model MASS)
- Total dustiness (from consideration of peak and period of plume) (model TOTDST)
- Total dustiness as a function of vehicle speed and mass of loose material on the road (model TOTDST/SPD)

By combining each of the aforementioned models, a preliminary estimate of dustiness associated with vehicle type, speed and wearing course was found, from which the required watering frequency (for water-spray based dust suppression) was determined. Table 4.11 summarises the independent variables used in the regression analyses. It should be noted that in applying the model described in Equation 4.17 to Equation 4.19, care should be taken when exceeding the range of parameter values used in deriving these models. Additionally, the time since last road maintenance was not included as an independent variable and steady-state conditions should be assumed when applying these Equations. A road that has just been

Table 4.11 Independent variables used in the dust defect models.

Independent variable	Description
$P_{0.425}$	Percentage of wearing course material passing the 0.425 mm sieve
SP	Shrinkage product, defined as;
	$LS \times P\,425$
	where LS = Bar linear shrinkage (%)
$P_{0.075}$	Percentage of wearing course material passing the 0.075 mm sieve
$LP_{0.425}$	Percentage of *loose* wearing course material passing the 0.425 mm sieve
PKDST	Peak dust reading ($\times 100$ mg/m^3) of the minus 10 micron dust fraction, measured by Hund Tyndalometer
TYPE	Indicator for truck type; 0 = Rear dump truck (RD) 1 = Bottom dump truck (BD)
WSHEAR	Wind shear (mm/s.mm) under the truck, defined as
	$$WSHEAR = \frac{V}{3.6 \times 10^{-3}.GRCLEAR}$$
	where V = Vehicle speed (km/h) GRCLEAR = Ground clearance (mm) under lowest part of vehicle
GVM	Gross vehicle mass (t) of fully laden haul truck
WHL	Number of wheels on truck
VOL	Hourly traffic repetitions on haul road

bladed, or an excessively ravelled or poorly performing road cannot be reliably modelled using this approach.

For the regression of the independent variables on mass of loose material on the road surface, the following model was given (after Thompson and Visser, 2002):

$$MASS = 4202,68 - 630,56.P425 + 1548,55.P075 + 78,75.SP$$
$$- 392,19.LP425 \qquad \text{Equation 4.19}$$

The model predicts an increase in the mass of loose material generated on the road when either the wearing course shrinkage product (representing plasticity and fines) or the percentage passing the 0.075 mm sieve increase. For the regression of the independent variables on total dustiness, derived from consideration of the peak dustiness (mg/m^3 minus 10 μm dust) recorded per vehicle pass, the following model was reported:

$$TOTDST = 2.92\ PKDST + 2260\ TYPE \qquad \text{Equation 4.20}$$

The absence of speed as an independent variable is partly explained by the peak values of dustiness measured increasing with increasing vehicle speed, at the lower test speeds the peak is low, but the period or duration is only slightly longer than at high speeds. This may in part be attributed to a slow vehicle generating dust only from the finer fractions of dust on the road; at higher speeds, the effects of wind shear, etc. entrain both small and large particles which tend to settle out faster than the smaller diameter particles entrained at low speed.

Finally, for the regression of the independent variables on total dustiness, derived from consideration of the mass of loose material on the road, traffic volumes and vehicle parameters, the following model was obtained:

$$TOTDST = (0.04\ MASS + 38.33\ WSHEAR + 0.12\ GVM\ WHEELS$$
$$- 2.44\ VOL)\ SPD + 2260\ TYPE \qquad \text{Equation 4.21}$$

The model predicts an increase in total dustiness with speed, mass of loose material on the road, wind shear (vehicles closer to the ground, travelling at higher speeds creating a higher wind shear effect), gross vehicle mass and the number of wheels. Traffic volume was negatively correlated with total dustiness, primarily due to the observation that higher traffic volumes led to a more compact wearing course, the removal of most loose material to the sides of the road and entrainment of spillage. This implies that although a high traffic-volume (busy) road may generate more dust per unit time than a low-volume road by virtue of the number of truck repetitions per unit time, the total dust concentration per vehicle pass will be lower.

Table 4.12 summarises the application of these models to a typical South African semi-arid (Highveld) surface mine operating rear-dump trucks on a well-built and maintained haul road. Using water-based spraying for dust suppression, the re-watering interval required in typical summer conditions is approximately 33 minutes to maintain a dust defect that at no time exceeds a score of two. Under winter conditions, this interval increases to approximately 63 minutes. When using bottom-dump trucks of a similar size and capacity, due to the increase in the number of wheels and the vehicle footprint area, greater concentrations of dust are more rapidly generated after watering and as such watering frequency increases to every 21 minutes in summer and 41 minutes in winter for the same traffic volumes and road wearing course material parameters.

The combinations of models previously described gave an insight into the required watering frequencies for various combinations of vehicle types, speeds, traffic volumes, wearing course material types and evaporation rates. This data can then be used as a base-case scenario with which to compare other types of dust palliatives under the same operating conditions.

Table 4.12 Model data and results for typical haul road watering frequency estimation.

Truck type	Rear dump (6-wheel)		Bottom dump (10-wheel)	
Season	**Summer**	**Winter**	**Summer**	**Winter**
Monthly evaporation rate (mm/mth)	**(260)**	**(140)**	**(260)**	**(140)**
Time to no effect (of palliation) (mins)	96.3	184.2	96.3	184.2
Percent palliation required[1]	65	77	65	77
Reapplication interval[2] (mins)	33	63	21	41

Notes
1 Average percent reduction in dustiness (compared with base-case untreated conditions) to maintain a dust defect score that at no time exceeds 23.5 mg/m³ –10 μm dust (equivalent to degree 2 defect score).
2 For a single application of 0.5 litres/m² water and assumes wearing course will not be adversely effected by water.

4.6.4 Palliative selection

In terms of the total ton-kilometres hauled on a mining haul road network, a significant number of the larger ton km operations have adopted various chemical dust palliation products with varying degrees of success. Many products are available which are claimed to reduce both dust and road maintenance requirements for mine roads. Often however, minimal specifications of their properties and no comprehensive comparable and controlled performance trials have been carried out in recognised, published field trials. Additionally, incorrect application techniques and construction methods often result, which leads to considerable scepticism about such products and their overall cost-effectiveness. In many instances on public unpaved roads, failures that could have been related to incorrect application, inappropriate management or unsuitable wearing course materials were often blamed on the product (Jones, 1999). However, when carefully chosen with regard to both the characteristics of the wearing course material itself, and the mine's road management philosophy and operating practice, results can be excellent.

From a mining perspective, the following parameters would define an acceptable dust palliative:

i.   spray-on application with deep penetration (the ability to penetrate compacted materials with generally low void ratios), or mix-in applications with minimal site preparation (rip, mix-in and recompact);
ii. straight-forward applications requiring minimal supervision, not sensitive nor requiring excessive maintenance or closely controlled frequent re-applications;
iii. the road should be trafficable within a maximum of 24 hours (short product curing period);
iv. availability in sufficient quantity at reasonable prices;
v.   adequate proven or guaranteed durability, efficiency and resistance to deterioration by leaching, evaporation, ultra-violet light and chemical reaction with wearing course or spillage on road;
vi. ability to test the product constitution to ensure consistency over time and deliveries;
vii. effective over both wet and dry seasons; and
viii. evaluated against local and international safety standards and environmentally acceptable.

From experience it was found that mix-in applications are far more effective than spray-on applications for an initial treatment. At least 100 mm of the wearing course should be treated as thinner layers result in delamination as a result of the shear stresses imparted by flexing of the tyres. A spray-on application, of say 1 litre/m^2 only results in a film thickness of 1 mm, which is hardly sufficient to resist the abrasive action of tyres.

The selection matrix in Table 4.13 can additionally be used to identify classes of palliative that would suit a certain application. However, the data does not specify the level of performance that could be expected, nor the average degree of palliation or degeneration rate expressed in terms of time from initial establishment and re-application rates. This information would be required as a precursor to an economic assessment of the selected palliative benchmarked against the base case of water-based spraying.

The broad classes of products available and their applicability are described in Table 4.14.

In road surfaces with too much gravel, dust palliatives do not appear to work effectively, more especially where a spray-on technique is used as opposed to a mix-in. The palliatives

Table 4.13 Selection guide for chemical palliatives.

	High PI (>10)	Medium PI (<10)	Sand	Wet weather trafficability	Ramp roads	Heavy traffic	Short term	Long term	Spray-on	Mix-in	Maintainable
Wetting Agents		☑			☑		☑		☑ I/R		☑
Hygroscopic Salts		☑				☑	☑		☑ R	☑ I	☑ M
Natural Polymer	☑	☑	☑				☑	☑		☑ I/R	☑ SO
Sulphonated Oils	☑	☑			☑	☑	☑		☑ I	☑ I	
Synthetic Polymer	☑	☑		☑	☑	☑		☑		☑ I	☑ SO
Tar/Bitumen Emulsions		☑	☑	☑	☑	☑		☑	☑ R	☑ I	☑ SR

Notes
☑ Suitable
I - Initial establishment application
R - Follow-on rejuvenation applications
M - Maintain when moist or lightly watered
SO - Maintain with spray-on re-application
SR - Maintain with spot repairs

do not aid compaction of the surface because of the poor size gradation, nor form a new stable surface. New surface area is created from exposed untreated material while, with a mix-in application, poor compaction leads to damage and ravelling of the wearing course; traffic induces breakdown of the material and eventual dust generation. With regard to water-soluble palliatives, rapid leaching may be problematic. In all cases, it is important to determine if the palliative can be applied with mine water (high TDS and/or salts), or if potable water is a requirement (as would be the case for some bituminous emulsion products where salt would 'break' the emulsion).

In compact sandy soils, polymer, acrylamide and bituminous-based emulsion products appear effective where leaching of water-soluble products may be problematic. However, in loose medium and fine sands, bearing capacity will not be adequate for the many products to maintain a new surface and degeneration will occur rapidly. In road surfaces with too much silt, it is unlikely that a dust suppression program will be effective. Excessive clay, silt or fine sand fractions may lead to a slippery road whilst poor bearing capacity leads to rutting and the need for road rehabilitation or maintenance, which destroys most products. Small-scale potholing has been observed on a number of roads following spray-on application or re-application, as a result of trafficking lifting fine cohesive material from the road. Again, where no depth of treatment has built up, this will lead to the creation of new untreated surface areas and the potential for potholes.

In general, spray-on applications do not appear appropriate for the establishment of dust treatments, especially with regard to depth of treatment required. A spray-on re-application or rejuvenation may be more appropriate, but only if penetration of the product into the road can be assured, otherwise it will only serve to treat loose material or spillage build-up, which

Table 4.14 Classes of dust palliatives and their performance.

	Hygroscopic Salts	Lignosulphonates	Petroleum- and tar-bitumen-based products	Others (Sulphonated petroleum, Ionic products, Polymers and Enzymes)
Climatic Limitations	Salts loose effectiveness in continual dry periods with low relative humidity. Selection dependent on relative humidity and potential to attract water into road surface. Minimum humidity for $MgCl_2$ is 32% and $CaCl_2$ is 70%.	Retains effectiveness during long dry periods with low humidity.	Generally effective, regardless of climate but will pothole (small diameter) in wet weather where fines content of wearing course is high.	Generally effective, regardless of climate.
Wearing Course Material Limitations	Recommended for use with moderate surface fines (max 10–20% < 0.075 mm). Not suitable for low fines materials or high shrinkage product or PI or a low CBR or slippery materials.	Recommended for use where high (<30% < 0.075 mm) fines exist in a dense graded gravel with no loose material.	Performs best with low fines content (<10% < 0.075 mm). Use low viscosity products on dense fine grained material, more viscous products on looser, open-textured material.	PI range 8–35 Fines limit 15–55% < 0.075 mm. Minimum density ratio 98% MDD (Mod). Performance may be dependent on clay mineralogy (enzymes).
Treatment Maintenance and Self-repair Capability	Reblade under moist conditions. $CaCl_2$ is more amenable to spray-on application. Low shrinkage product materials may shear and corrugate with high speed trucks. Shear can self-repair.	Best applied as an initial mix-in and quality of construction important. Low shrinkage product materials may shear and corrugate with high speed trucks. Tendency to shear or form 'biscuit' layer in dry weather – not self-repairing.	Requires sound base and attention to compaction moisture content. Slow speed, tight radius turning will cause shearing – not self-repairing, but amenable to spot repairs.	Mix-in application – sensitive to construction quality. Difficult to maintain – rework. Generally no problem once cured.

Tendency to Leach out or Accumulate	Leaches down or out of pavement especially in wet season. Repeated applications accumulate.	Leaches in rain if not sufficiently cured. Gradually oxidise and leach out. Repeated applications accumulate.	Does not leach. Repeated applications accumulate.	Efficacy depends on the cation exchange capacity of the host material. Repeated applications accumulate.
Comments	A high fines content may become slippery when wet. Corrosion problems may result.	Generally ineffective if wearing course contains little fine material or there is excessive loose gravel on the road.	Long lasting – more effective in dry climates. May cause layering after several spray-on re-treatments – especially where fines content. >15%<0.075 mm.	Generally ineffective if material is low in fines content or where loose gravel exists on surface. Curing period may be required.

will rapidly breakdown and create new untreated surfaces, and layering can occur, (as shown in Fig. 4.36) where the build-up of treated fines on the surface leads to a smooth slippery surface devoid of any of the original aggregate in the wearing course. A spray-on treatment is however useful to suppress dust emissions from the untrafficked roadsides, since it would be easier (and cheaper) to apply and, with the material typically being uncompacted, would provide some depth of penetration and a reduction in dust emissions from truck induced turbulence. Since the treatment forms a skin, no vehicular or pedestrian traffic should be permitted, else the protection is damaged.

For chemical-based dust suppressants, the average degree of dust palliation and the period over which it is applied can be considerably better than that achievable by water-based spraying alone. However, in terms of cost-effectiveness, an evaluation is required with which to determine the extent of the cost benefits attributable to chemical-based dust suppression, together with an indication of those factors likely to alter the trade-off between water- and chemical-based dust palliations.

4.6.5 Economic evaluation of palliation strategies

A pre-requisite of any cost evaluation is a model that provides a rapid means of making a consistent comparison of the real costs of alternative control measures. For dust control, this requires an analysis of the relative costs of alternative palliation options, such that the most cost-efficient option can be determined, together with an indication of the sensitivity of the selection in terms of the primary modelling parameters. Changes in cost of dust control resulting from the introduction of alternative strategies are utilised to evaluate dust control options. This allows the economic implications of the introduction of alternative strategies to be expressed in terms of a base-case cost, often the use of water-based spraying.

The development of a model requires identification of the key components that affect the overall cost of dust control and their interrelationship and effect on the total cost of road treatment (costs/m²). The major cost elements considered in dust control include the following:

Figure 4.36 Delamination of a spray-on treatment.

- Capital equipment
- Equipment operating costs
- Road maintenance frequency
- Material cost (palliative cost)
- Activity-related costs and efficiencies, such as surface preparation, dust palliative application, grading, watering, compaction and finishing, for either a mix-in or spray-on establishment or rejuvenation re-applications

These parameters are, in turn, influenced by the selected palliative, application methodology and frequency for the specified level of control, as shown schematically in Figure 4.37.

The benefits derived from the application of any palliative must be completely characterised to fully determine the value of the dust suppression system. The primary benefit, that of improved air quality and road- and driver-safety, is problematic to assess. It has been established that a truck driver's exposure to the dust produced, and for how long, is not critical for the characteristic dust involved, except where a combination of open-cabs and short-hauls or high traffic volumes are found (Thompson and Visser, 2003a). The safety benefit is more tangible – the current practice of applying water at frequent intervals is often insufficient (especially in hot and dry climates with high evaporation rates), in view of the low levels of dustiness and distant visibility required to be maintained on a mine haul road. This would imply that by reducing the frequency of watering, through the use of palliatives, this may offset the additional costs of material and construction required for their effective application. Table 4.15 summarises various primary data classes recognised in the analysis in terms of cost or benefit, highlighting those used in developing the cost model described by Thompson and Visser (2006b);

Palliative performance is assessed in terms of the product's establishment application rate (l/m^2), establishment method (spray-on or mix-in) and re-application rate (l/m^2 spray-on only) and frequency, to achieve a comparable average degree of dust palliation as to that achieved by water under the same conditions. Palliative establishment and rejuvenation rates and methods reported in Table 4.16 are tailored to the individual application and its constraints, which, where different from those indicated, may change the rates and frequencies quoted. The performance achieved may therefore not represent the optimal performance of the palliative selected. Care should be taken when specifying the modelled or alternate establishment and rejuvenation strategies – it is important that the data used closely reflects the anticipated palliative performance under the (often unique) conditions of its application. Applying palliatives that result in semi-permanent surfacings may also require a change in the conventional motor-grader maintenance to innovative procedures such as collecting dust with a truck-mounted industrial vacuum cleaner or brooming the road. Consequent capex and associated costs would have to be considered.

Figure 4.38 summarises a site-specific analysis for several of the palliatives discussed previously, using a maximum allowable dust defect score of 2 or 4. Details of the palliative cost, establishment and re-application rates and methodologies, together with the average degree of palliative performance achieved for the particular combination of haul road, wearing course and traffic volumes used in this example are presented in more detail by Thompson and Visser (2000a).

In this particular case, as the allowable dust defect reduces from 4 to 2, any palliation method will become more expensive to apply. Using a cost-index referenced to the cost of watering the haul road at establishment (Year 0) to limit dust defect to ≤4, as the required

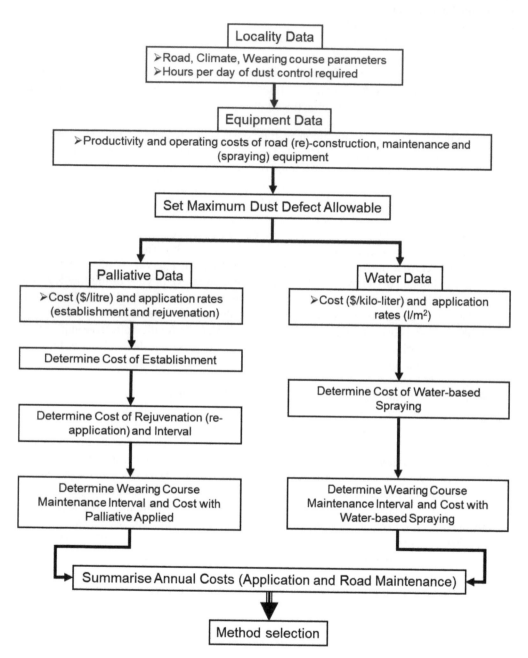

Figure 4.37 Schematic model of dust palliative cost evaluation.

level of control increases, so does the cost index. However, at these low levels of control, water-based palliation remains cost-effective. However, in the case of control at ≤2, polymer and tar/bitumen emulsion products show a cost benefit over water-based spraying, the polymer emulsion only following establishment after year one (initial costs being higher due

Table 4.15 Cost/benefit considerations.

Benefit	Included in model	Cost	Included in model
Improved safety	X[1]	Surface preparation, palliative establishment application and finishing costs (equipment and material)	✓
Improved health	X[1]	Palliative re-application (equipment and material)	✓
Reduction in grading cost	✓	Remaining grading costs	✓
Reduction in grading frequency or gravel loss	✓	Remaining watering costs	✓
Reduction in watering cost	✓[2]	Reduced safety (cost of accidents)	X[1]
Reduction in vehicle down-time and maintenance	X	Reduced health (cost of exposure to low air qualities)	X[1]
Improved hauler cycle times	X[3]	Reduced water-cart fleet utilisation	X[2]

Notes
1 Not analysed, but if model comparison is based on the cost to achieve a specified level of control efficiency, many of the costs become equal or do not apply. For example, the comparative safety and health benefits from reduced dustiness would become equal, irrespective of control methodology applied (water-spray base-case or palliative).
2 Reduced utilisation of water-cart or road maintenance fleet would not necessarily generate savings, except for reduced maintenance, parts and fuel, since vehicle and driver would still be required. However, where the fleet consists of more than 1 vehicle in use, the reduction in numbers may generate savings by parking up a vehicle.
3 Improved hauler cycle times are significantly affected by rolling resistance. Even relatively small reductions in rolling resistance on a laden haul can be beneficial to cycle time and tonnes/hour.

Table 4.16 Typical palliative application and performance modelling estimates.

Palliative	Application methodology (l/m^2 @ interval) (S = spray-on, M = mix-in) for specified dust defect score			
			Dust defect score 2	Dust defect score 4
Water	Spray @ 0.5 l/m^2	S	Every 33 mins	Every 68 mins
Hygroscopic	Establishment	S	2 l/m^2	2 l/m^2
salts	Re-application	S	0.5 l/m^2 every 10 days	0.25 l/m^2 every 12 days
Ligno	Establishment	M	1 l/m^2?	1 l/m^2?
sulphonates	Re-application	S	0.2 l/m^2 every 10 days	0.1 l/m^2 every 14 days
Petroleum	Establishment	M	1.2 l/m^2	Unknown
resins	Re-application	S	0.11 l/m^2 every 20 days	Unknown
Polymer	Establishment	M	0.95 l/m^2	0.95 l/m^2
emulsions	Re-application	SS	0.05 l/m^2 every 15 days	0.025 l/m^2 every 20 days
Tar/bitumen	Establishment	M	3 l/m^2	3 l/m^2
emulsions	Re-application	S	0.125 l/m^2 over 15 days	0.063 l/m^2 over 15 days

to higher activity and product-related establishment costs), whilst the tar/bitumen-emulsion class of product is cheaper throughout the life of the road, by virtue of lower establishment and re-application quantities and costs. In this example, there is no water consumption cost, only the operational (spraying) costs are included. If a cost for water is applied, this would

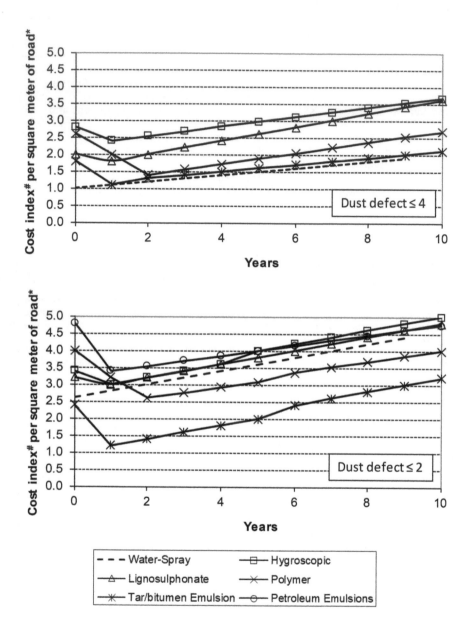

Index based on water-spray palliation costs at establishment (Year 0) and control at dust defect ≤ 4.

* For particular mine site wearing course material and unit operating costs

Figure 4.38 Economic evaluation of dust palliatives for two levels of dust control.

therefore significantly enhance the cost-effectiveness of using dust palliatives at both the control levels specified.

The economic evaluation can be extended to explore the most cost-effective option in the case of roads where the existing wearing course does not meet the minimum specifications introduced earlier. In this case, four options typically exist, namely water-based spraying or chemical palliation, both in conjunction with or without improvement to the existing wearing course material. The cost effectiveness of the various management strategies for either water-based spraying, the application of a chemical palliative, or improvement of the wearing course material, can be analysed as outlined previously. This is achieved through consideration of the establishment and reapplication costs associated with each type of wearing course and palliative combination, in comparison to the additional costs associated with the placement of the improved wearing course. In this latter case, the activity cost of wearing course improvement is inclusive of palliative establishment, thus reducing costs of establishment whilst leveraging the benefits of an improved wearing course, in terms of reduced palliative application frequencies and rejuvenation rates.

In all cases, it is important to consider road management philosophy, in particular, to be sure that the use of palliatives or stabilisers will result in reduced road deterioration rates, less maintenance interventions and hence maintain a lower overall rolling resistance. It is obviously counter-productive to use palliatives whose performance is better and longer than that of the road on which they are applied – since the palliative will be destroyed when maintenance is carried out to fix road-generated deterioration issues.

4.6.6 Practical evaluation of additives

A major concern is the suitability of an additive when applied for the first time in an environment. The South African Agrément Council for example has developed a certification procedure for non-traditional stabilisers. It is based on strength and erosion resistance of four standard materials. Non-traditional stabilisers are products for which no standard tests exist, such as polymers, enzymes and sulphonated oils. The question that remains is the applicability of these results in a specific environment of a mine.

Generally, the additive supplier would be requested to provide material for at least 500 m of haul road. Constructing a road of this length is not a trivial exercise and also has significant cost implications. Invariably only one or two products can be evaluated because of the length of road with uniform characteristics required. What is of even greater importance is the complication if the product does not perform as anticipated and the road has to be closed to remove the wearing course.

For this reason, a small-scale evaluation procedure was developed (Visser and Erasmus, 2005). A panel of 3 m × 1 m × 150 mm thick is constructed with a combination of wearing course material and additive, and even additive dosage, on any road to evaluate the performance under traffic. Suppliers would be willing to donate the required quantity of additive, and construction can be done labour intensively by a maintenance gang mixing the wearing course and the additive (Fig. 4.39(a)). Compaction is done by a 1 ton walk-behind vibratory roller in two layers 75 mm thick (Fig. 4.39(b)). Testing is performed after 28 days to allow the mix to cure, and the test is performed at in-situ moisture content and after soaking for two hours under a head of 50 mm of water. Large sieves or cut oil drums

Figure 4.39 Preparation and testing of small-scale panels.

(Fig. 4.39(c)) provide the confinement for the water. The DCP is used to measure the in-situ strength (Fig. 4.39(d)).

As an example five panels were constructed to evaluate four different additives (Products A to D), together with a control or untreated section. The CBR results obtained with the DCP are given in Table 4.17. As expected the CBR values are much higher at in-situ moisture

Table 4.17 In-situ strength results after 28 days.

	At in-Situ Moisture Content		Soaked for 2 Hours	
	CBR(%)	*Layer depth (mm)*	*CBR(%)*	*Layer depth (mm)*
Untreated	100	150 mm*	19	95
			50	50
Product A	>200	40 mm	43	200
		120 mm		
Product B	130	65 mm	17	280
	29	120 mm		
Product C	>200	40 mm	100	150
	110	50 mm		
	60	100 mm		
Product D	45	140 mm	13	150

* CBR (100) over layer depth (150 mm)

Source: Visser and Erasmus, 2005.

content than after soaking. The value of this small-scale experiment was that Products B and D had no influence on strength and in fact had a lower support value than the untreated material. This was because the additive required clay to interact, and the material was sandy. Product C had a significant benefit at both moisture contents. By means of this experiment is was possible to eliminate two products as being unsuitable.

Chapter 5

Road management and maintenance

5.1 Background and orientation

Design and construction costs for the majority of haul roads represent only a small proportion of the total operating and road maintenance costs. The use of an appropriate road maintenance management strategy has the potential to generate significant cost savings – particularly in the light of increases in rolling resistance due to the interactive effects of traffic volume and wearing course deterioration. With large trucks being used, it is inevitable that some deterioration or damage to the road will occur, and this damage needs to be regularly repaired. The better the road is built, the slower the rate of deterioration and thus the less maintenance required. A poor road, however, will quickly deteriorate and will need frequent maintenance (often to the detriment of other roads in the network). Poor roads can result both from inappropriate design, or a good design, but with inappropriate construction techniques and quality control.

This chapter details at the various approaches and techniques of road maintenance, its impact on costs of material haulage and methods that can be applied to optimise haul road maintenance and minimise total road user costs.

5.2 Introduction to road management

The management of mine haul roads has not been widely reported in the literature, primarily due to the subjective and localised nature of operator experience and required road functionality levels. In most cases, comment is restricted to the various functions comprising road maintenance, as opposed to the management of maintenance to minimise overall total costs. Some rules of thumb imply adequate serviceability (functionality) can be achieved by the use of one motor grader (and water car) for every 45,000 tkm of daily haulage. The United States Bureau of Mines Minerals Health and Safety Technology Division (USDOI, 1981) in their report on mine haul road safety hazards confirm these specifications, but without a clear statement as to what activities comprise road maintenance. Other approaches (Hawkey, 1982; Long, 1968) include blading the road after every 90 truck passes (based on largest RDT operating at those times). What is clear from this is that road performance varies significantly as does the material types comprising the wearing course. In Chapter 4 it was shown how the combination of poor wearing course material selection, coupled with high traffic volumes, can lead to a rapid loss of functionality and the generation of road defects that contribute to an increase in rolling resistance. At a mine investigated it was found that a poor wearing course material resulted in a high rolling resistance, and even when additional road maintenance resources were applied, the problem remained; only when the wearing course was corrected was productivity restored and conventional road maintenance re-established. High rolling resistance and any over-riding safety considerations should have the greatest influence on what type of road maintenance solution is applied and when.

5.2.1 Root cause analysis in road maintenance

Before road maintenance management is introduced, it is worthwhile to consider why road maintenance is carried out in the first place: its primary purpose is to restore the road to its original operating specification, i.e. to conserve the integrity of the road wearing course by returning or redistributing the gravel surface and removing defects. In most cases, this will improve a road and reduce its rolling resistance to a more acceptable 2–2.5%. How quickly the road deteriorates again (i.e. rolling resistance increases) will dictate when the next maintenance activity occurs. All too often though, road maintenance is done with little recognition of the following:

i.   **where** the maintenance was done (what road segment of the network); and
ii. **what** was done (blading, dragging, shallow rip and re-grade or re-sheet, etc.).

Keeping road maintenance records of where and what are important, since this information can indicate whether or not a road is performing well, and if not, what the problem is. The approach is similar to a Root Cause Analysis (RCA) – identify why a segment of road is maintenance intensive before any repair work is scheduled. The maintenance report can also be cross checked with any other information from the dispatch system (if any) to validate and confirm the problematic zones against truck speed along the roads, suspension strut pressure spikes, excessive temperature on tyres, operators reports, etc.

Take, for example, the defects shown in Figures 5.1 and 5.2 – a fairly large area of sinkage, squeezing or potholing on the road. No amount of grading will 'fix' this problem since, as was explained earlier, these problems indicate failure deeper in the road layerworks and simply to recover and spread wearing course material into the depression will not cure the root cause of the problem. Once the root cause is recognised (structural failure), one can plan to fix the problem correctly (box out to remove softs and backfill and compact with selected base-layer material, re-establish the wearing course and compact).

Several more examples are shown in Table 5.1 and are described to assist in the process of 'reading' a road and determining the root-cause of either poor performance or frequent maintenance interventions on the road.

It is important to remember that a RCA for mine road 'failures' is just as valid as it would be for any other asset. Excessive maintenance on poorly performing segments of the road network is symptomatic of some underlying design issue. When 'reading' the road, work through each design component and question whether or not that component is correct before moving to the next potential design shortcoming. In that way, the root cause of the under-performance can be isolated in a structured manner and appropriate solutions planned, scheduled and implemented.

5.2.2 Road maintenance activities

What exactly is 'road maintenance'? There are several key activities that encompass road maintenance, from routine or 'patrol' road maintenance (blading or grading), through to resurfacing, rehabilitation and betterment, as defined in Table 5.2. However, the ideal maintenance

Figure 5.1 Road defect of sinkage, squeezing or potholing.

Figure 5.2 Road defect of sinkage, squeezing or potholing.

Table 5.1 Examples of road defects and underlying 'root-causes'.

There is plenty of loose, unbound material on the road. It will require frequent blading, due to instability. But also consider the geometry here – note how the junction is on grade. Many of the problems here are associated with accelerating laden trucks from stand-still on grade and the high wheel torque, which shears the wearing course. Root cause is both the current wearing course itself (Sp too low), but primarily the poor junction geometric design.

Geometry is most likely the issue here – the crown of the road is non-existent. Water drains to the centre of the road. But also consider structure – maybe a soft spot under the road has resulted in this deformation. If that were the case (and DCP probing could confirm this), it would be necessary to remove the softs, backfill and compact selected waste rock and re-establish wearing course.

When a road looks like this after blading, one has a build-up of fines (clay, mud) on the surface and the grader is simply spreading this around, or if this material persists to depth (100–200 mm) then the wearing course is nowhere near specification. Root cause is lack of coarse fractions (up to 40 mm) in the wearing course – if it is a spillage issue, deep ripping, remixing and water (to OMC) and recompact the wearing course will bring the material back to specification (if the spillage build-up is deep, blade it off the road and remove first) to expose 'original' wearing course.

Two issues apparent here, first, tyre tracks in the road indicate a wearing course material either poorly compacted or, in its compacted state, failing to reach the minimum 80% CBR required. Second extensive rutting is also seen (depressions in the road in the wheel paths of the truck), indicating too soft a structure to support the wearing course. The root cause here would be primarily structural – even the 'best' wearing course will not perform well if the underlying support is poor.

The edge of the road is in poor condition as the road structure was not constructed as wide as the surfacing. It could quite simply be solved by moving the road boundary markers back onto the edge of the constructed road – if the operating truck width would allow for this. The root cause here is that the road was built for smaller trucks and now that larger trucks are in use, the road boundary markers have been moved out to accommodate 3.5x the width of the largest truck. But the construction width of layerworks (geometric design) does not extend this far so edge failure will occur.

There is oversize which may be either 'base' layer dump rock material showing through a too-thin wearing course layer – or a poorly sized wearing course itself. Blinding the top of the base layer may assist to prevent these protrusions. If that is not the issue then the wearing course may be too thin (due to erosion, routine maintenance or normal gravel loss) and a resurfacing is due.

strategy for mine haul roads should be the one that results in the minimum total cost since the mine must both pay to maintain and to use the road. Two elements form the basis of total costs, namely road maintenance costs and vehicle operating costs (VOC). Both these cost elements are directly related to road condition or more specifically pavement roughness progression – commonly referred to as rolling resistance.

The selection of a maintenance management program for mine haul roads should be based on the optimisation of these costs, such that total vehicle operating and road maintenance costs are minimised. Current operating practice however does not always reflect this maxim and Figure 5.3 shows a typical hierarchy of road management systems, from the low-complexity ad-hoc approach, to fully cost-optimised real-time management system. An optimal road design will include a certain frequency of maintenance (grading, etc.) within the limits of required road performance and minimum vehicle operating and road maintenance costs. In mine road maintenance, it is the translation of this maintenance frequency concept into a maintenance strategy that is problematic. The major difficulty encountered when analysing maintenance requirements for haul roads is the subjective and localised nature of the problem: levels of functionality or serviceability being user-, site- and material-specific.

5.2.3 Routine haul road maintenance – untreated wearing courses

Irrespective of the maintenance management system used, there are some key elements of 'good practice' of mine road routine maintenance that should be observed for either type of unoptimised management approach:

- Ad-hoc maintenance – reactionary maintenance management in response to poor haul road functionality. Typically managed by daily inspection of the road network and a subjective assessment of road segment functionality and maintenance priorities.

Table 5.2 Types of road maintenance activities.

Mode	Activity	Effect
Routine (Patrol) Maintenance	Spot regravelling	Fill potholes and depressions and exclude water
	Drainage and verge maintenance	Reduce erosion and material loss, improve roadside drainage
	Dragging/Retrieval	Redistribute surface gravel
	Shallow blading	Fill minor depressions and ruts and reduce rolling resistance
	Dust control/watering	Reduces loss of binder and generation of dust
Resurfacing	Full regravelling	Restore thickness of wearing course.
	Deep blading	Reprofile road and reduce roughness. Remix wearing course material provided that the wearing course has at least 100 mm of gravel.
Rehabilitation	Rip, regravel, recompact	Improve, strengthen or salvage deficient pavement.
Betterment	Rehabilitation and geometric improvement	Improve geometric alignment and structural strength.

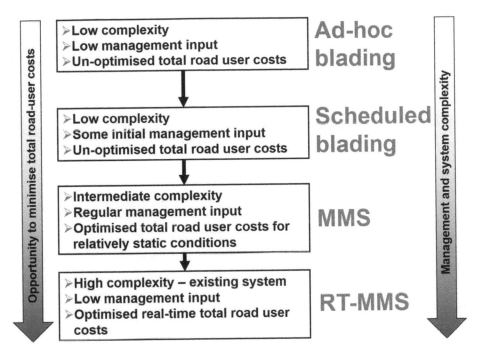

Figure 5.3 Hierarchy of haul road management systems.

- Scheduled maintenance – road network is maintained according to a fixed schedule or frequency, irrespective of the actual functionality of the road segment being worked.

Routine or patrol grading with motor-grader is critical to the performance of haul roads, since the use of unpaved and unbound material will result in some inevitable deterioration under the action of traffic, as discussed in Chapter 4. The basic purpose of grading is to maintain a satisfactory running surface, free of any extensive defects, and to keep the road well drained. Grading can have a relatively short-term effect especially with poorer types of material and reductions in rolling resistance will be only very short-term. Improper grading usually lowers the surface and eventually, unless material is periodically bladed in from the roadside, the base material is exposed, which would weaken the overall structure (due to moisture ingress) and could potentially damage truck tyres if a selected blasted waste rock base is used (this type of base, however, would not be susceptible to water ingress).

Large motor graders (4.3 to 7.3 m blade width) are typically used for road maintenance: light ripping, scarifying, opening and mixing and finishing of the wearing course. As a general rule the size of a grader should match the size of the trucks used on the haul roads to optimise the productivity of the grader considering the road width and the number of blade passes necessary to maintain the road (Fig. 5.4). Since the moldboard is usually angled when moving material, an effective blade length must be computed to account for this angle. This is the actual width of material swept by the moldboard (Effective Blade Length).

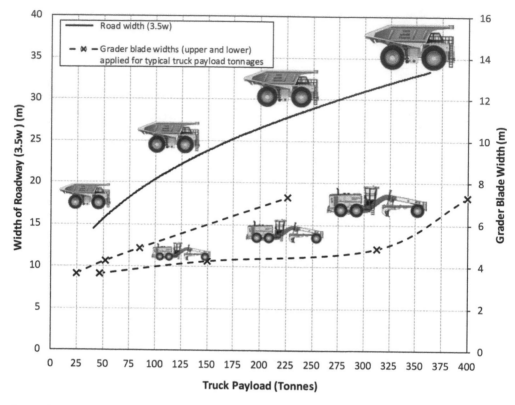

Figure 5.4 Matching the size of grader with the truck size and road width.

Good grading practice consists of lightly watering the haul road to be maintained (or maintaining after rain) and then conducting the following:

- Winning and recovery of wearing course passes
- Removing defects, shaping and mixing passes
- Spreading passes

This process is illustrated in Figure 5.5 in which the width of overlap is generally 0.6 m (2.0 ft.). This overlap accounts for the need to keep the tyres out of the windrow on the return pass. Figure 5.5 shows bringing wearing material in from the sides (winning and recovery passes) or cutting down high sections of the surface and filling the low spots with the surplus loose wearing course material. However, this should not be at the expense of loss of profile (cross-fall, crown or super-elevation, etc.). Where the profile of the road has been lost, Figure 5.6 shows the process of winning material from roadside drains/shoulders to restore the profile (camber or crown).

Ideally, prior to grading, water should be applied to the road surface if it had not rained. If the formation is damp the loose material graded into low spots may be compacted by traffic to give a more uniform surface with little loss of material. In dry weather maintenance

Winning and recovering material passes (drains and shoulders)

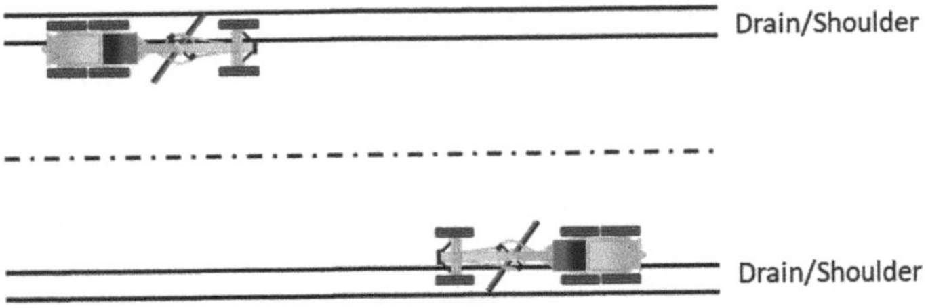

Drain/Shoulder

Drain/Shoulder

Blades angled as close as possible to 45° to recover/win wearing course from drain and shoulder. First and second pass apply where a crowned profile is maintained. For a cross-fall across width of road, win only from the drain side. In all cases avoid blading spillage on to the road.

Cutting road surface, removing corrugations and defects and mixing

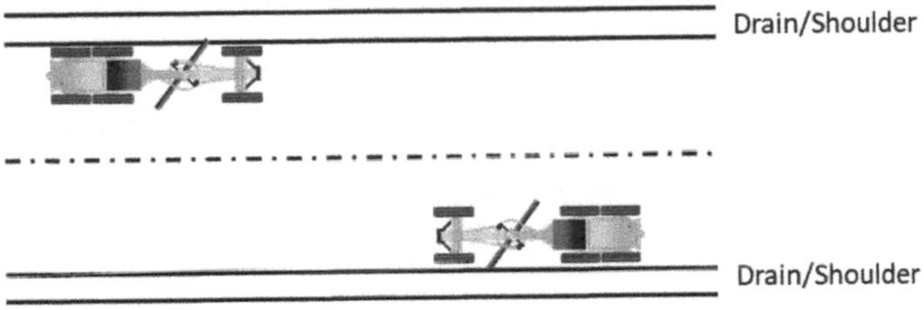

Drain/Shoulder

Drain/Shoulder

Blade hard on surface and scarify if blade not cutting to bottom of corrugations or other defects. Multiple passes in each direction may be required to reach the centre of the road from both sides.

Spreading and finishing

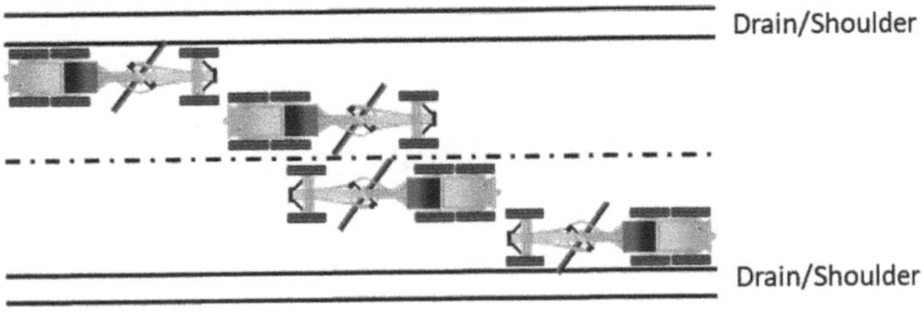

Drain/Shoulder

Drain/Shoulder

Two or more spreading passes will be required depending on the amount of material obtained. If a crowned profile is to be maintained, the blade must not extend beyond the centre of the road. When a cross-fall profile is adopted, blading can be applied in successive passes across the full width of the road, working up from the drain side

Figure 5.5 Routine grading practices. (Care should be taken not to blade spillage back into the road).

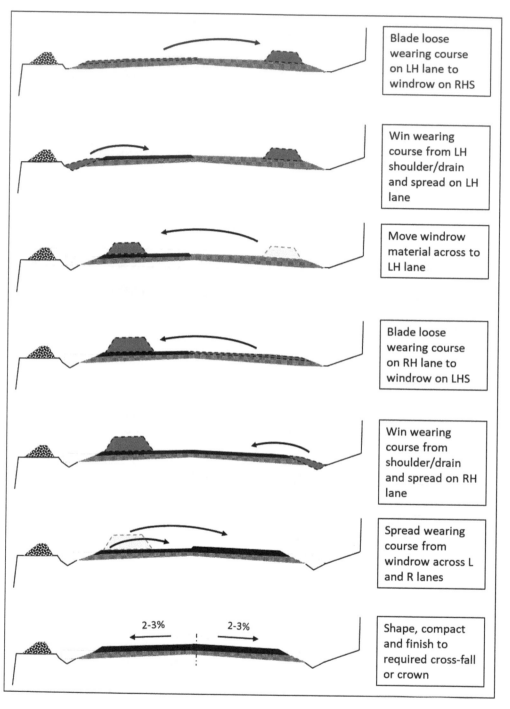

Figure 5.6 Grader passes required to win wearing course from shoulder and drains to re-establish profile of road.

grading should be confined to smoothing loose material on the surface, keeping the cutting of high sections to a minimum and especially avoiding breaking up the hard crust. Major grading operations should be confined to times when the road can be left un-trafficked and the surface material damped down by the application of water and preferably compacted.

The grading process should blade loose or wet material (after rainfall) from the outside edge to the centre of a crowned road (common practice is to blade to the edge of the road – but this can interfere with drainage). This material, when inspected and dried, should not contain excessive fines if it is intended for re-spreading and compaction when dried to OMC. If the loose or wet material contains excessive fines, it should be removed from the road and either screened to remove fines before being considered for resurfacing, or replaced as per wearing course specifications (typically with the addition of a 'bonier' aggregate to counteract the impact of the higher fines material content).

Active roads should be inspected to determine the extent of defects and whether a routine maintenance intervention is required – in particular, road profiles should be maintained at or near their original design profiles. Active roads should also be inspected after heavy rainfall events to isolate areas of scour or water damage that require repair prior to trafficking. The section of road being graded should be confined to the length that can be fully completed in a predominantly dry period over a complete shift. Some basic road maintenance activities often carried out include the following:

- Scarify or top up the wearing course surface material to bring coarse fresh material to the surface – especially when the surface slicks up since surface spillage (mud, etc.) will change the characteristics of the wearing course if left on the road. This is often seen as laminations (layering) of fine material.
- Remove spillage and other loose material from the road surface – do not leave as windrows.
- Repair areas that have slumped or settled or wet spots by backfilling with base or wearing course as appropriate.
- Maintain drainage on road and shoulders as per specification (keep roadside drains open).
- Maintain or re-establish cross fall/crown and super-elevation profiles.

On the first pass of the grader, the moldboard top-edge should lead the toe (cutting) by 50 to 100 mm until material characteristics are evident. The moldboard orientation should enhance the cutting and material 'rolling' action, as this is when productivity of the equipment is maximised. Bulldozing with the blade on the other hand is slower and less productive.

Shallow holes may be repaired without prior treatment by being cleared of loose material and filled with properly selected fresh (or recovered) wearing course. Both the road itself and the added material should be moist at the time, or should be moistened shortly after filling the hole. Patches should be finished to a smooth surface by grading and compaction. Larger defects on the road surface can sometimes be restored by loosening and reshaping. This can only be achieved if there is sufficient thickness of wearing course material and the quality of the layerworks material has not been compromised.

If the road surface is adequately designed and maintained, larger repairs will usually be required only at isolated points. Where a depression is too large to be eliminated by grading, it should be ripped, boxed out and filled with material similar to that in the adjoining road (selected blasted waste rock as per layerworks specifications) and/or wearing course (as per wearing course specifications), depending on the depth, and finished to a smooth surface by grading and compacted.

If depressions or large potholes are caused by subgrade failure, the road should be repaired by excavating back (boxed out) to at least twice the depth of the base layer design thickness, the sides of the excavation being cut vertically. It will be necessary also to excavate subgrade material, replacing the full depth with selected blasted waste rock as per layerworks specifications. Wearing course (as per wearing course specifications), should be placed, finished to a smooth surface by grading and compacted. However, large depressions at the shoulder of a road, especially in fill sections, should be thoroughly investigated to preclude potential fill slope failure as a source of the problem. Where benches daylight to natural ground, the bench edges could be unstable, particularly if spoil or fill has intentionally or unintentionally been side-cast, and/or bench drainage does not prevent run-off at low points.

When finishing the haul road, the moldboard top-edge should lead the toe (cutting) by 150 to 250 mm to maintain a cutting edge approximately normal to the cut surface.

Resurfacing the road will eventually be required to restore the thickness or specification of the wearing course and should be done:

i.   at the first indication of significant distress under traffic (a larger number of defects which require intensive maintenance activities);
ii.   before the base layer begins to protrude through the wearing course (i.e. at less than 100 mm thickness); and
iii. if the wearing course material has weathered and eroded such that it no longer meets functionality requirements.

5.2.4 Routine haul road maintenance – treated (palliated) wearing courses

In Chapter 4, the road management philosophy was identified as an important factor to consider to be sure that the use of palliatives or stabilisers will result in reduced road deterioration rates, less maintenance interventions and hence maintain a lower average rolling resistance. It is obviously counter-productive to use palliatives whose performance exceeds that of the road on which they are applied – since the palliative will be destroyed when maintenance is carried out to fix road structure deterioration issues.

Therefore, the routine maintenance requirements of a palliated or stabilised wearing course are significantly different from a conventional wearing course and greater emphasis is placed on road 'cleaning' as opposed to road defect removal and gravel redistribution. Spillage and especially fines imported onto the road surface from truck tyres needs to be removed and often, the road surface itself would be treated again, with a spray-on maintenance (re-) application to bind any remaining loose particles on the surface.

However, in doing this it is important to note that the fundamental characteristics of the wearing course will change as a greater proportion of fines become embedded on the surface of the road and as a result, functionality, especially wet- and dry-skid resistance, will deteriorate, along with the creation of thin laminations made up of layers of fine treated material.

Under certain circumstances, especially where a significant depth of treatment exists (from a mix-in establishment application), the surface may be lightly scarified with a toothed blade (as opposed to the usual flat blade). The Sandvik System 2000 Grader Edges are an example, illustrated in Figure 5.7. Toothed or scarified blades attached to the grader

Figure 5.7 Example of toothed grader blade (L) and resultant road surface finish (R).

Source: image [LHS] courtesy Greg Kostiniuk Inc.

moldboard work best with the tilt of the moldboard set back to 70°. If the tilt angle is not correct, chattering or vibration will occur. This treatment is useful to remove minor functional defects whilst at the same time not totally destroying the hard-running surface. Where extensive defects (in terms of degree and extent) are seen in a treated wearing course, remediation would involve significant reshaping and this would in all likelihood destroy the treatment. Under these circumstances, the root cause of the poor functional performance should be addressed prior to re-treating the road, since in blading the road, the treatment is destroyed along with any economic benefit it delivers.

This shows that there is a need to repair defects such as potholes, depressions, corrugations and rutting by means of motor-grader interventions. However, a haul road with a semi-permanent surfacing should be able to withstand damage by tyres if properly constructed. Some mines are now moving towards long-life roads where a minimum of grader maintenance would be required. Under those circumstances the main defect would be dust from spillage or carry-on by tyres. A motor-grader would not be efficient in removing dust, and cutting the hard surface has to be avoided. The maintenance strategy has to be adapted, and often a rotary broom shown in Figure 5.8 is used. Unfortunately a significant cloud of dust is generated (Fig. 5.8) and, in the absence of wind, the dust will again settle on the road, where an unacceptable dust cloud would be created by the haul trucks. In some operations water is sprayed by the broom operation, which reduces the dust and a roll of mud is then swept off the road. This approach may not be highly effective. An alternative procedure would be to use a truck-mounted vacuum cleaner, as is used on some diamond mines, to collect the dust

Figure 5.8 Rotary broom and dust cloud generated by brooming.

Source: image LHS www.forconstructionpros.com/equipment/earthmoving/product/10077682/champion-motor-graders-acquired-by-volvo-motor-grader-attachments.

without abrasion. The collected dust would then be disposed of in a waste dump. Spillage of rocks would have to be removed manually or by a 'hit-squad', and trucks should be loaded to the correct capacity to avoid rock spillage. It must be appreciated that using semi-permanent surfacings would require a change in the mine road management strategy.

5.2.5 Grader selection and road maintenance productivity

The motor grader is used in a variety of applications in a variety of construction sectors. Therefore, there are many ways to measure its operating capacity or production. One method expresses a motor grader's production in relation to the area covered by the moldboard:

$$A = S\left(L_e - L_o\right)1000\,E_{ff}$$

Equation 5.1

Where

A: Hourly operating area (m²/h)

S: Operating speed (km/h). Figure 5.9 shows a chart with typical speeds for different activities for a typical 4.267 m (14′) moldboard grader

L_e: Effective blade length (m). Figure 5.10 shows the effective blade lengths for typical moldboard settings of 60° and 45°. When the moldboard is normal to the direction of travel, productivity is maximised, but care needs to be taken to maintain material rolling action. Moldboard angles of 60–80° are normally used in light, free-flowing material. Lower moldboard angles of 45–60° are required when working heavy cohesive material, mixing materials and with or drain work. If material flows around the leading edge, further reduce the blade angle

L_o: Width of overlap (m)

E_{ff}: Job efficiency – Job efficiencies vary based on job conditions, activity carried out and operator skill, etc. A good estimation for efficiency is approximately 0.70–0.85, but actual operating conditions should be used to determine the most appropriate value. Higher values can be assumed if the task comprises longer passes with fewer turnarounds and an overall better road condition (functionality), requiring fewer passes to remediate the road.

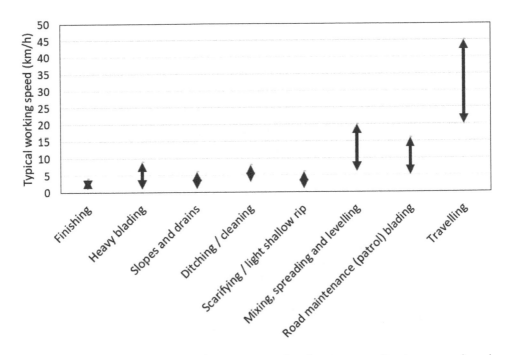

Figure 5.9 Operating speed ranges for various road maintenance applications – road grader.

Moldbord Length (m)	Effective Length (@60° blade angle) (m)	Effective Length (@45° blade angle) (m)
3.66	3.17	2.59
3.96	3.43	2.80
4.27	3.70	3.02
4.87	4.22	3.44
7.31	6.33	5.17

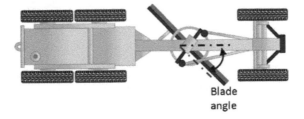

Blade angle

Figure 5.10 Determination of effective blade length.

An alternative approach is described by Visser (1981) in which road grader productivity is reduced with increasing road roughness, modified by Thompson and Visser (2006a) based on observation and time study data from various surface mining operations. Grader productivity was theoretically calculated based on the area of road maintained per hour, per metre effective pass width (L_e-L_o). Typically 6000 m^2/h/m blade width was found to be in broad agreement with operational practice.

Most grader operators envisage an increase in the number of blade passes required to achieve an acceptable finish when a road had significantly deteriorated (associated with a higher rolling resistance and functional defect score). A productivity curve was proposed, incorporating this reduction in grader efficiency associated with excessively rough roads as shown in Figure 5.11. The grader productivity factor from this figure is multiplied by the typical 6000 m^2/h/m blade width.

5.2.6 Maintenance management systems (MMS)

Although various road maintenance strategies can be applied depending on the type of mining and complexity of the operation, ideally, road-user costs should be minimised and road performance maximised. This requires a more systematic approach to road maintenance, referred to as a maintenance management system (MMS). This approach considers how quickly roads deteriorate under the action of traffic and how deterioration impacts vehicle operating costs, compared with how much it would cost to maintain the road (both costs being carried by the mine or transport contractor) measured against the benefits of an improved road performance. Using an ad-hoc or even a routine-based maintenance management system will not deliver minimum total costs.

The concept is shown in Figure 5.12, where total costs comprise vehicle operating costs (VOC) and road maintenance equipment and application costs. Clearly, when the frequency of maintenance increases, a better overall functionality is maintained and, in terms of defects, the corresponding rolling resistance would be lower. Although it would be more expensive

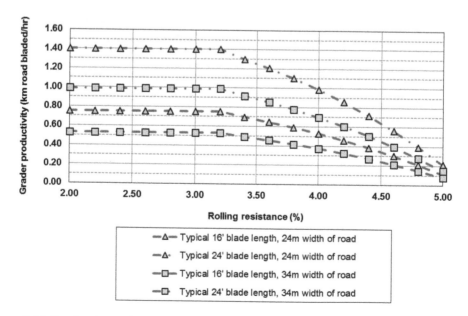

Figure 5.11 Productivity of a motor-grader during routine haul road maintenance operations.

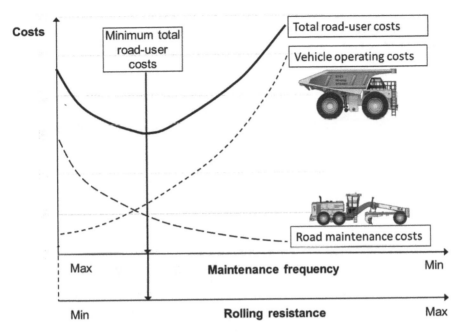

Figure 5.12 Concept of MMS – determining the optimal frequency of road maintenance commensurate with minimum total road-user costs.

to apply very frequent maintenance to the road, it would be advantageous to truck costs through a lower overall rolling resistance, which would reduce fuel burn, cycle times, etc.

However, not all haul roads, or segments in a network of mine haul roads, will exhibit the same costs profile as shown in Figure 5.12. The key to minimising total cost across a network of road segments in a mine is to determine which segments have the greatest influence on cost per tonne hauled. Typically long, flat, high-traffic volume hauls should enjoy priority in any maintenance system – since a small reduction in rolling resistance will have the greatest influence on reducing cost per tonne hauled across the network. That priority should also be reflected in the resources applied to design these roads in the first instance. In comparison, shorter (steeper) and/or lower traffic volume roads will have less of an impact on total costs across the network, since although rolling resistance may be at a higher value than other roads, the penalty associated with this increased rolling resistance, in terms of total costs, would only be small and the costs associated with maintenance will not necessarily be recovered through reduced cost per tonne hauled.

Mine haul road maintenance intervals are closely associated with traffic volumes, operators electing to forgo maintenance on some sections of a road network in favour of others. This implies an implicit recognition of the need to optimise limited road maintenance resources to provide the greatest overall benefit. This optimisation approach is inherent in the structure of the MMS for mine haul roads. Two elements form the basis of the economic evaluation, namely these:

i.   haul road functional performance – rolling resistance-based model of deterioration; and
ii.   vehicle operating (VOC) and road maintenance cost models.

MMS is designed for a network of mine haul roads, as opposed to a single road analysis. For a number of road segments of differing wearing course material, traffic volumes and speed and geometric (grade and width) characteristics, together with user-specified road maintenance and vehicle operating costs, the MMS approach can be used to determine the following:

i.   the maintenance quantities as required by the particular maintenance strategy;
ii. the vehicle operating and road maintenance costs; and
iii. the optimal maintenance frequency for segments of the network such that total road-user costs are minimised.

This aspect of road management is discussed further in Section 5.4, once the impacts of poor road performance have been described from the perspectives of truck performance and cycle time and the development of vehicle operating costs models described.

5.3 Minimising total costs across a network of roads

Central to the cost of truck hauling is the concept of resistance (expressed here as a percentage of Gross Vehicle Mass [GVM]). Rolling resistance is expressed in terms of kg (or N) resistance per ton of GVM, where approximately 10 kg/t = 1% rolling resistance or 1% grade (against the direction of travel). Rolling resistance is defined as the force required to maintain a vehicle at a steady speed on level ground and is a function not only of the gross vehicle mass and drive-train characteristics, but also of the type and condition of the tyres and the road surface on which the vehicle is operated (together with other smaller additional resistances, such as wind resistance, etc.). Simply stated, the truck engine has to supply more power to overcome these additional resistances and the greater these resistances are, the more power is needed, which means higher fuel or energy consumption, or lower attainable speeds.

Whilst rolling resistance is to some extent a controllable variable in a mining operation, grade resistance is more typically fixed and is a function of mine geometry and ore: waste stripping ratios, etc. As discussed in Chapter 2, each truck has an optimal grade at which travel time is minimised between the limits of

i.   a long-haul distance, but high-speed, low-gradient haul route; and
ii. a shorter-haul distance, but low-speed, higher-gradient haul route.

Similar to rolling resistance, grade resistance is a measure of the gravitational force that must be overcome to move a truck against the (unfavourable uphill) grade. Likewise, for favourable downhill grades, grade 'assistance' is defined. Grade is measured in percent slope (ratio between vertical rise or fall and the horizontal distance) and the resistance (or assistance) is approximated as the grade percentage of GVM.

As was discussed in Chapter 1, increases in resistance can have a significant impact on truck cycle time, effectively slowing the vehicle due to the additional resistance to motion. This effect will also be seen as an increase in fuel burn and ultimately cost per tonne material hauled. In a large and deep surface mining operation, truck haulage costs can account for up to 50% of the total operating costs incurred by a the mine (as shown in Fig. 5.13) and any savings generated from improved road design and management benefit the mining company directly as reduced cost per tonne of material hauled.

Surface Mining Costs
Open pit mining 250kt/day 500m depth

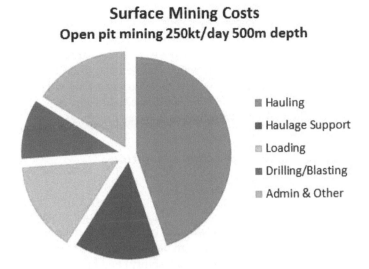

- ■ Hauling
- ■ Haulage Support
- ▥ Loading
- ■ Drilling/Blasting
- ▥ Admin & Other

Figure 5.13 Typical cost breakdown for a deep open-pit operation.

The in-pit loading benches in any mining operation are generally the poorest running surfaces in an operation and although only representing a small fraction of the overall haul cycle, they can significantly influence costs – not so much from the perspective of cycle time or fuel burn, but rather as a result of the potential for tyre or truck frame damage. The use of impact rollers (as discussed in Chapter 3) is a useful technique to prepare the running surface of these benches where the use of waste rock as a blind or temporary running surface would not be feasible (e.g. ore benches). Similarly, at waste dump tip heads, poor surface conditions are common and aggressive, albeit over a limited distance, but given the truck manoeuvring that is common in these areas (loading bench and tip heads alike) they should not be over-looked when it comes to road management. 'Temporary' roads, in what ever form, do permanent damage to a truck using them.

In Chapter 1 it was shown how rolling resistance impacts truck cycle time and fuel burn, an additional 1% ramp road rolling resistance will reduce truck speed (and increase travel time) by 8–11%, whilst on a flatter surface road a rolling resistance increase from 2% to 3% will reduce attainable truck speed by between 18% and 23%. In this section, the source of this rolling resistance effect will be analysed with reference to the haul truck speed-rimpull-gradeability characteristics.

Rolling resistance estimates include an empirical assessment based on tyre penetration that typically specify a 0.6% increase in rolling resistance per centimetre tyre penetration into the road, over and above the 1.5% (radial and dual wheel assemblies) to 2% (cross-ply or single wheel assemblies) minimum resistance. Thus a 2 cm tyre penetration would equate to a rolling resistance of 2.7% for a rear dual assembly. In using this approach however, note should be taken of the following:

- Tyre 'penetration' in this context can also equate to road surface flexure, but by using the structural design approach described in Chapter 3, displacement under the tyre will be minimised and rolling resistance approach the minimum for the road condition.
- Tyre inflation pressures and tyre tread patterns can also influence rolling resistance.

Table 5.3 summarises other empirically derived values that would apply to unbound gravel pavements. These approaches however do not specifically link wearing course performance to rolling resistance and also fail to account for the effect of traffic volumes on road deterioration.

Chapter 4 presented an approach to rolling resistance evaluation, which linked the type of road wearing course material and road functional defects, to an equivalent in terms of rolling resistance. This approach also considers the effect of traffic volumes and the rate of increase in deterioration (expressed as rolling resistance) and is the approach adopted in the following sections when an estimate of rolling resistance is required.

5.3.1 Haul truck speed and cycle time estimation

5.3.1.1 Speed rimpull gradeability curves

The truck speed-rimpull – gradeability curve is supplied by truck manufacturers to enable estimations of truck cycle time to be determined based on the relationship between mass of the vehicle, engine rimpull and effective grade (defined in Chapter 1). Typical performance charts for propulsion (rimpull-speed-gradeability) and braking (retard) are shown in Figure 5.14 for a selected model of electric-drive truck (diesel engine driving electric rear-wheel motors via a generator) and in Figure 5.15 for a selected model of mechanical drive truck (diesel engine driving rear wheel via torque converter, gearbox and drive shafts). These types of charts, irrespective of the vehicle power train (electric or mechanical) can be used to evaluate the impact of rolling resistance on truck performance, to derive a more accurate estimation of the rules of thumb previously mentioned.

To use these charts in developing an estimate of the impact of rolling resistance on truck performance, the following basic steps can be followed:

- Determine which curve on the chart to use – where effective grade is against the load (uphill), propulsion curves are read, whereas if effective grade is with the load (downhill), then the retard curve is used.
- As an example, for a segment of a haul cycle – Ramp1 is built at a grade of 8% and has road rolling resistance of 2% related to the truck speed. If the truck is working laden against the grade, the truck speed can be determined by

Table 5.3 Typical rolling resistance estimates for unbound gravel road surfaces.

Rolling resistance (%)	Road surface conditions (built from unbound gravel materials)
2	Strong layerworks and hard, compacted (stabilised) well-built and maintained road, no tyre penetration/deflection discernible
2–3	Intermediate strength layerworks, compacted (stabilised), well-built and frequently maintained road, with minimal (<25 mm) tyre penetration/deflection
3–5	Weak layerworks or surfacing material, 25–50 mm tyre penetration/deflection, rutted and poorly maintained
5–8	Weak layerworks or surfacing material, 50–100 mm tyre penetration/deflection, rutted and poorly maintained

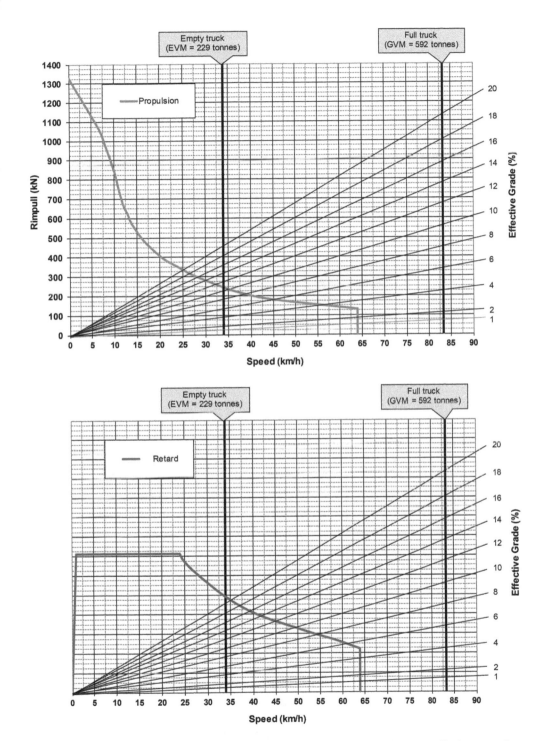

Figure 5.14 Examples of truck speed-rimpull-gradeability and braking (retard) charts – electric drive truck.

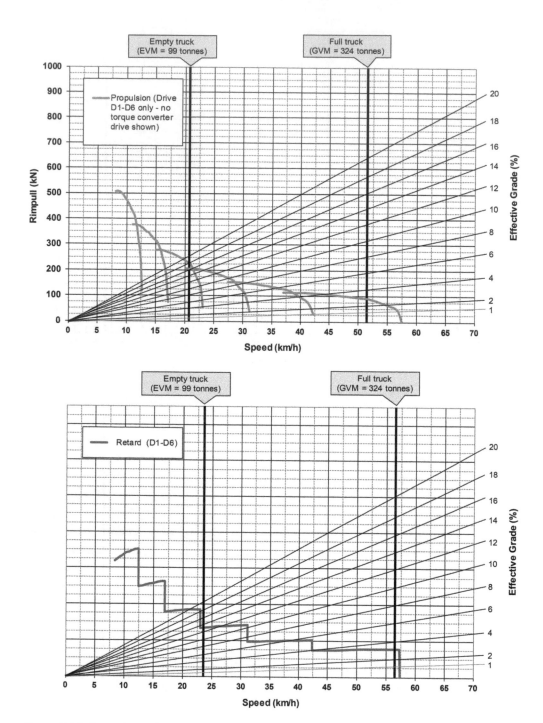

Figure 5.15 Examples of truck speed-rimpull-gradeability and braking (retard) charts – mechanical drive truck.

i.   calculating effective grade (8%+2%) = 10%;
ii. selecting the 10% effective grade line, follow this line diagonally to its intersection with (in this case) the vertical line representing the full truck;
iii. using a straight edge, project this point horizontally to the left until it intersects with the propulsion curve; and
iv. using a straight edge, project this intersection point vertically down to read the speed of the truck.

Using the 10% effective grade against the load and the propulsion curve in Figure 5.15, a truck speed of 15 km/h is found. Using the same logic, but in this case for the empty truck return journey at an effective grade of (8% − 2%) = 6%, the maximum truck speed is 57 km/h from the intersection of the 6% effective grade line with the empty truck vertical line, projected (right) to the retard speed curve (whether or not it is safe to travel down a 8% physical ramp at this speed has been examined in Chapter 2). Figure 5.16 shows the process applied.

Table 5.4 shows the earlier example applied now to a hypothetical haul consisting of two segments and is used to show how cycle time could be impacted as a result of an increase in rolling resistance (from 2% to 3%), based on two segments of haul; Ramp1 (8% grade) and ex-pit (1% grade) roads, both of a 2000 m length. In this case an increase of 11.7% is seen in the travel components of the cycle time, due the decrease in truck speed associated with the 1% increase in road rolling resistance.

The approach explained here can also be extended to plot the speed-gradeability characteristics of a given truck (at a specific GVM), as shown in Figure 5.17. From this type of data, the incremental speed losses associated with a 1% increase in resistance can be plotted, as shown in Figure 5.18. In this specific (truck model) case, using Figure 5.16, at a laden truck speed of 49.7 km/h (at 3% effective grade), a 1% increase in rolling resistance equates to an effective grade of 4% at which the speed becomes 37.1 km/h, a reduction of 25%. The magnitude of the percentage speed losses gradually reduces with increasing grade of road, such that a 1% change in rolling resistance at 10% grade equates to a 9% speed reduction. Note that for electric drive trucks, the speed reduction curve is smoother and does not exhibit the peaks and valleys seen with mechanical drive trucks. However, typically similar magnitudes of percent speed losses are seen, above 20% for flatter roads, decreasing to typically 10% on steeper ramp roads.

5.3.1.2 Coopers speed estimation models

Cooper (2008) developed a number of equations to quickly determine truck speed without recourse to individual truck speed-rimpull-gradeability curves. These enable a rapid approximation of truck speed to be made and, as will be seen in Section 5.3.2, also enable estimates of fuel consumption to be determined as input into a vehicle operating cost model.

Two equations for truck speed are proposed, the up-ramp (unfavourable, against the grade) truck speed (v_{uf}, km/h), dependent on the energy available from the engine applied to the truck weight, including its payload, to overcome rolling resistance, transmission losses and load elevation requirements. The down-ramp (favourable with the grade) truck speed (v_{fav}, km/h) is dependent on the energy (generated by the truck running downhill with gravity, less rolling resistance and transmission losses) that the engine can absorb, either through braking or retard/blower system capacity.

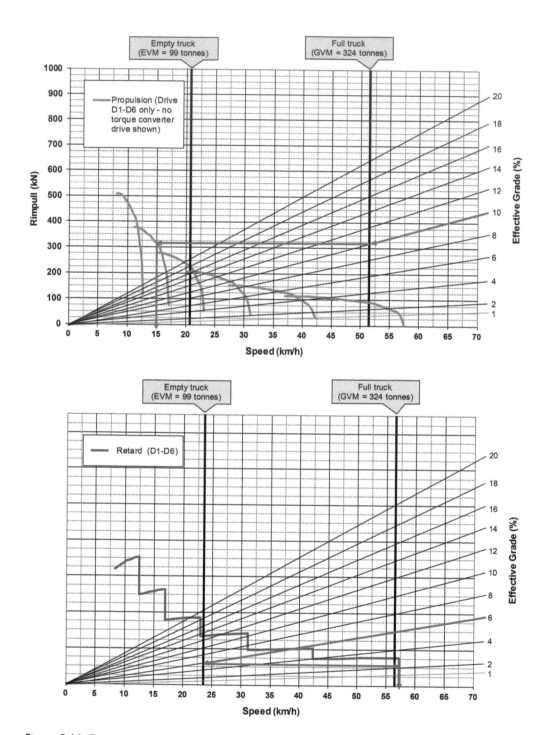

Figure 5.16 Determining truck speed from speed-rimpull-gradeability and braking performance characteristics.

Table 5.4 Calculation of cycle time increase due to 1% increase in rolling resistance.

Segment	Grade %	RR%	Effective Grade %	Distance (m)	Speed (km/h)	Travel time (s)	Percent decrease in speed for 1% increase in RR
Laden Unfavourable Grade							
Ramp1	8	2	10	2000	15.0	480	
Ex-pit	1	2	3	2000	49.7	145	
Unladen Favourable Grade							
Ex-pit	1	2	1	2000	57.0	126	
Ramp1	8	2	6	2000	57.0	126	
Total cycle time (travel) (s)						878	
Laden Unfavourable Grade							
Ramp1	8	3	11	2000	13.5	533	10
Ex-pit	1	3	4	2000	37.0	195	26
Unladen Favourable Grade							
Ex-pit	1	3	2	2000	57.0	126	0
Ramp1	8	3	5	2000	57.0	126	0
Total cycle time (travel) (s)						981	
Percent increase in travel time due to 1% increase in RR							**11.7%**

For truck operating against the grade (up-ramp):

$$v_{uf} = \frac{0.367\,\epsilon_t P_{Gross}\sqrt{(1+GR^2)}}{GVM\,(GR+RR)}$$

Equation 5.2

For trucks operating with the grade (down-ramp):

$$v_{fav} = \frac{-0.367\,\epsilon_r P_{Gross}\sqrt{(1+GR^2)}}{GVM\,(GR+RR)}$$

Equation 5.3

Where

P_{Gross} = Gross engine power (SAE J1995) (kW)
ϵ_t or ϵ_r = Transmission or retarder efficiency
GVM = Vehicle mass (t)
GR = Grade of road (-ve downgrade, +ve upgrade) (%)
RR = Rolling resistance of road (%)

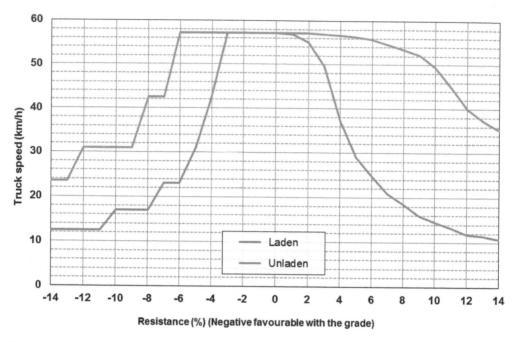

Figure 5.17 Speed-gradeability characteristic of the truck performance curve defined in Figure 5.15.

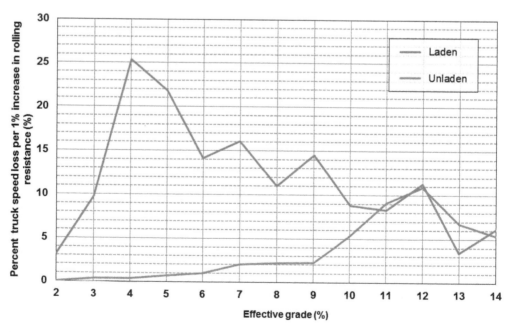

Figure 5.18 Effect of rolling resistance on percent loss of speed.

The transmission and retarder efficiency factors are used to calibrate the model to the manufacturer's performance curves. Cooper reports a transmission efficiency of 0.8 and a retarder factor of 1.15 appears to work well universally. The model will work for electric trucks and trolley assist applications but calibration of the transmission efficiency (electric power direct to rear wheel motors) may be required. In other atypical applications such as down-grade laden hauling, some trucks will have additional retarder capacity options fitted in which case the retarder factor would have to be increased to model laden haulage favourable (with) the grade.

Using the previous speed-rimpull-gradeability curves (Fig. 5.15) and the manufacturer data outlined above in conjunction with Equation 5.2 and Equation 5.3, Figure 5.19 shows how the estimation equation compares to the manufacturer data. Although there is some overestimation from the typical unladen downgrade speeds, laden speeds are well modelled, and these are the most critical variants in the estimation of cycle time. As will also be seen later, in terms of fuel consumption in a vehicle operating costs model, downgrade segments are generally run with the truck engine in retard and as such the effects of these estimation errors are small from the perspective of fuel burn.

5.3.1.3 Spline estimation techniques

Thompson (2017) applied a one-way (monotonic) spline to estimate truck speeds, following an approach described by Perdomo (2001) and later Erarslan (2005). Unlike polynomial or other curve estimation techniques, a spline is forced to directly pass through all points in a data set. The one-way spline is a constrained version of the Bessel spline which always produces monotonic results as long as the source data is monotonic, which typically applies to the speed-rimpull curves of mine haul trucks (SRS1 Software, 2017).

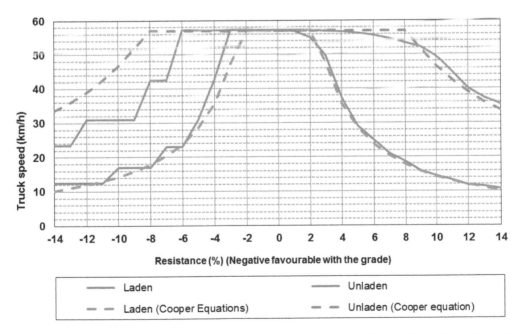

Figure 5.19 Estimation characteristics of Cooper truck speed equations for selected mechanical drive truck (Fig. 5.15).

From the speed-rimpull-gradeability curve of a particular model of truck, engine power can be determined at each point on the curve from:

$$PR = \frac{v}{3.6RP}$$

Equation 5.4

Where

PR = Power delivered as rimpull (kW)
v = Truck speed (km/h)
RP = Rimpull (kN)

The transmission efficiency (ϵ_t) at each point can be determined as:

$$\epsilon_t = \frac{PR}{P_{Gross}}$$

Equation 5.5

Where

P_{Gross} = Gross engine power (SAE J1995) (kW)

The speed reduction ratio at each point (SRR) can be determined as:

$$SRR = \frac{2\pi\epsilon_t RPM_{Max}}{60\dfrac{v}{3.6}}$$

Equation 5.6

Where

RPM_{Max} = Maximum engine RPM

And the drive reduction ratio (DRR) as:

$$DRR = \frac{2\pi a RPM_{Max}}{60\dfrac{v_{\epsilon tmax}}{3.6}}$$

Equation 5.7

Where

V_{etmax} = Truck speed at maximum efficiency (ϵ_{tmax})
a = Static loaded radius of truck tyre (m)

Thus, at each point on the speed-rimpull curve, engine RPM can be found from:

$$RPM = \frac{DRR\,60\dfrac{v}{3.6}}{2\pi a}$$

Equation 5.8

And power developed at each point is thus:

$$P = \frac{RP\dfrac{v}{3.6}}{\epsilon_{tmax}}$$

Equation 5.9

A monotonic spline is then applied to the engine power and RPM data from which rim-pull can be found at any point on the curve from torque (T, kNm):

$$T = \frac{P}{\dfrac{2\pi RPM}{60}}$$

Equation 5.10

And the associated rimpull as:

$$RP = \frac{T\,DRR\epsilon_{tmax}}{ar}$$

Equation 5.11

Using this approach, a truck speed-rimpull-gradeability chart can be constructed and interrogated for a number of variable operating parameters including engine power, wheel size (radius), gear (if mechanical drive truck), weight of vehicle and grade or effective grade, to determine speed, rimpull and rolling resistance as appropriate. Figure 5.20 illustrates the full spline estimation model to accommodate any of the aforementioned variables and to predict the associated truck speed.

ROLLING RESISTANCE ESTIMATION FOR CAT793F											
From AEHQ6868-10 Engine CAT C175-16 For Truck KT02 Engine Serial Number B12B34567											

ENGINE and TYRE DATA				OPERATING DATA				CALCULATED			
Gross Actual Engine Power kW	RPM @ Gross Actual Engine Power	Operating Limit RPM	Wheel radius (m)	Gear (D1-D6)	RPM	Mass (tonnes)	Grade of road (%)	Speed (km/h)	Rimpull (kN)	RR %	TR %
SAE J1995	SAE J1995	Limiting RPM	Static loaded radius	Range 1-6 (No TC)	OK	Range 100-400t	Range 0-16%				
2001	1750	1923	1.59	D2	1598	390	9.7	13.61	438.92	1.83	11.48

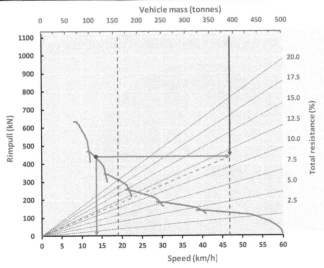

Figure 5.20 Reconstruction of 'live' speed-rimpull – gradeability curve from cubic spline interpolation (in this case back-calculating rolling resistance from truck operating point).

5.3.2 Vehicle operating cost models

Vehicle operating costs vary with rolling resistance and the vehicle operating cost (VOC) model refers to how truck operating costs change incrementally with changes in road rolling resistance. A typical VOC cost breakdown is shown in Figure 5.21, from which it is clear that the contribution from fuel, tyres and repairs and truck maintenance is over 70% of the total costs (due in part to the depth of the workings and duty of the trucks). Thus a VOC model should consider the effect of increasing rolling resistance on fuel consumption, tyres and vehicle wear and repair.

The prediction of fuel consumption variation with road rolling resistance is central to any MMS and fuel consumption itself is a significant component of total vehicle operating costs. Fuel consumption of vehicles on public roads has been shown to vary primarily with vehicle type and speed, and the total resistance of the road (Chesher and Harrison, 1987) and a similar approach is used to estimate fuel consumption for mine haul trucks.

Using the estimation equations developed by Cooper (2008), fuel consumption can be determined with reasonable accuracy (within 10% as reported by Cooper). However, in this application where the incremental cost penalty or advantage associated with changes in rolling resistance is sought, it is not so much the actual value determined that is important, rather the increment in that value. In any MMS model it is the change of a particular cost item with increasing road roughness which is of major concern as opposed to a fixed cost, although the latter is important in assessing the relative contribution of that cost to total costs.

Haul truck fuel consumption (Q_f) (litres/hour) is based on the engine load (D_{Load}) (% of gross power available). Engine load is developed when the effective grade is greater than 0%:

$$D_{load} = \frac{2.72\, v_{uf} GVM \left(GR + RR\right)}{P_{Gross}\, \epsilon_t \sqrt{\left(1 + GR^2\right)}} \qquad\qquad \text{Equation 5.12}$$

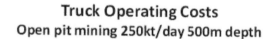

Truck Operating Costs
Open pit mining 250kt/day 500m depth

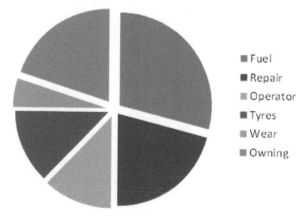

- ■ Fuel
- ■ Repair
- ■ Operator
- ■ Tyres
- ■ Wear
- ■ Owning

Figure 5.21 Typical cost breakdown for haul trucks operating in a large, deep open-pit operation

Where

v_{uf} = Truck speed against the grade (km/h)
P_{Gross} = Gross engine power (SAE J1995) (kW)
ϵ_t or ϵ_r = Transmission or retarder efficiency
GVM = Vehicle mass (t)
GR = Grade of road (%)
RR = Rolling resistance of road (%)

Fuel consumption is then found from:

$$Q_f = 0.226\, P_{Gross} \left(D_{load} + 0.03\right)$$ Equation 5.13

Applying the equations above to the truck model depicted in Figures 5.15 and 5.19 results in a fuel consumption profile for the truck as shown in Figure 5.22.

5.3.3 Road maintenance cost models

Road maintenance costs are based on the grader productivity characteristics described in Section 5.2.5 and Figure 5.11 and the unit operating costs, together with an estimate for water-cart productivity and associated operating costs (since ideally, a light water spray should be applied to the road prior to maintenance blading). When a smaller machine is applied to a larger than ideal road width, then productivity would fall as the number of blade passes required to complete the road would increase. For water cart productivity, figures quoted (Thompson, 1996; Thompson and Visser, 2006a) are approximately 12,500 m²/h/m of the width of the spray pass.

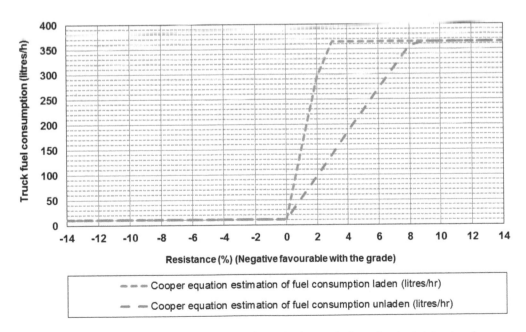

Figure 5.22 Fuel consumption variation with grade for typical mechanical drive truck.

5.3.4 Tyre cost models

In the analysis of tyre costs for large haul trucks a number of problems exist relating to the applicability of the data available on which such models can be based. Since a mine has a limited number of roads of variable quality, any model of cost variation with road condition or other geometric parameters will not be particularly robust. Other limitations exist with regard to damage attributable to loading or dumping areas as opposed to the road itself; up to 80% of all large tyres fail before they wear out. Cuts are responsible for about 45% of failures, with impacts causing nearly 30% (Caterpillar, 2017; Ingle, 1991). This would obscure any road condition effect on tyre costs. It is also necessary to consider the effects of tyre TKPH (tonne-kilometre per hour), compound selection (higher TKPH, higher tyre wear rates, lower TKPH, lower wear rates) and the abrasive effects of the various types of material found in the wearing course. All these additional factors make it difficult to determine a suitable model; suffice to say that a poorer quality road must ultimately reduce tyre life when compared to a better quality road. In the absence of definitive data, recourse can be made to an underlying hypothesis of a road condition and geometry-related tyre cost relationship. Such a model (after Thompson, 1996) is of the form:

$$TW = \left(62.52RR - 29.31 + 2|GR|\right)10^{-3}$$

Equation 5.14

Where
TW = Tyre set wear rate (tyres consumed per 1000 km for a six-wheeled truck)
GR = Value of road grade (%) as the positive value is used
RR = Rolling resistance (%)

At 2% rolling resistance on a flat haul, the model predicts a tyre set wear rate of 0.1 tyre sets/1000 km (0.016 tyres per 1000 km), or 62,500 km per tyre. Given an average truck speed of 12 km/h this equates to approximately 5200 hrs tyre life. At higher grades, this life reduces to about 4000 hrs per tyre. When rolling resistance increases by 50%, tyre consumption increases by 65% and road geometry effects on tyre consumption equates to a 12% increase in consumption for a change in grade from flat-haul to 10% (higher rolling resistance roads showing a reduced grade wear contribution). No curvature effects were modelled since this effect is generally assumed to be insignificant for large trucks, especially if speed limits are applied to tight radius curves.

5.3.5 Vehicle maintenance, parts and labour cost models

Vehicle maintenance and repair costs comprise both the cost of the parts consumed and the labour hours expended on the repair and maintenance of the vehicle. These costs are related to the type of vehicle, its age, how the vehicle is used – especially the haul route characteristics. This cost component of the total vehicle operating cost has been shown to be a significant contributor to the benefits from road improvements (Paterson, 1987) when considering public roads. From the perspective of mine roads, however, the contribution is significantly less due to the predominance of the fuel consumption costs discussed previously.

Similar data limitations exist with respect to individual mine parts and labour cost data as with tyre data, with additional complications of costs not being easily ascribed to a particular vehicle type where more than one vehicle type is used for hauling and the influence of high cost long-life replacement parts fitted (especially with mixed-age fleets). The analysis, interpretation and transferability of any model will be dependent on individual mine maintenance strategies, truck speeds, duty, driver behaviour, the level of preventative maintenance and the history and average age of the vehicles in the fleet. It may be anticipated that, across mines, differences exist in policy and expenditure on maintenance which should ideally be addressed by developing a site-specific model for these costs.

With these data limitations in mind, recourse was made to established models to provide a suitable point of departure in estimating suitable models for parts and labour costs. The common practice of road user cost studies has been to express vehicle parts consumption in terms of a standard parts cost. This represents the parts consumption as a fraction of the replacement price of the vehicle and for mine trucks, Thompson and Visser (2006a) estimated the model as:

$$\frac{PC}{VEH_R} = (92.28RR - 128.78) H^{0.375} \qquad \text{Equation 5.15}$$

Where
PC = Parts cost (/1000 km)
VEH_R = Replacement cost of vehicle ($\times 10^{-6}$)
H = Vehicle age khrs (total engine hours)

On roads with an average 2% rolling resistance, the model predicts a standardised parts cost per 1000 km of 116 at 5000 engine hours. Taking a replacement cost of 10 million, this equates to a parts cost of 1.16/km. When average rolling resistance increases by 50%, standardised parts cost increases to 292 and when engine hours double to 10 khrs, standardised parts costs increase by 30%.

In the estimation of labour costs relating to increased maintenance labour quantities per unit distance to parts consumption per unit distance and the effect of rolling resistance, a basic model, Eq. 5.15, was proposed by Thompson and Visser (2006a) such that:

$$LC = \alpha \left(\frac{PC}{VEH_R} \right)^\beta \qquad \text{Equation 5.16}$$

Where
LC = Labour costs (/1000 km)
α, β = Constants representing labour unit costs and the proportion of parts costs associated with significant maintenance labour activities

The coefficient β is often less than unity, varying from 0.47 to 0.65 (for buses and trucks). Increases in parts costs are predicted to lead to an increase in labour costs but at a decreasing rate which may reflect the relatively capital-intensive nature of major repairs on large haul trucks and their modular construction. Engines, wheel motors, etc. are often removed as a complete unit to be repaired off-site, the only labour cost being recorded arising as a result of removal and replacement of items as opposed to their repair. Thompson (1996) adopted a value of 0.45, based on limited mine site data.

The coefficient α is reported to be affected by road condition in some (commercial truck) studies, both increasing the labour cost (with fixed parts cost), suggesting that maintenance activities for vehicles on poorer haul roads are relatively more labour intensive, and decreasing the labour cost implying less labour at a given parts cost being applicable. In the case of mine haul trucks, due to their unitised construction it may be anticipated that no additional road condition effects will be present in this cost item, other than that appearing as an explanatory variable in the standardised parts cost.

Although not specifically included, depreciation, or recovery of the purchase price through use, is dependent on operating conditions. Table 5.5 shows the significant impact of application severity on truck life (articulated dump trucks in this case) after Caterpillar (2017). For RDTs, these hours would be much larger.

5.3.6 Optimising management of mine roads

Mine haul road maintenance frequency is closely associated with traffic volumes, operators electing to forgo maintenance on some sections of a road network in favour of others. This implies an implicit recognition of the need to optimise limited road maintenance resources to provide the greatest overall benefit. This optimisation approach is inherent in the structure of the maintenance management system (MMS) for mine haul roads. Two elements form the basis of the economic evaluation, namely these:

i.   haul road functional performance – rolling resistance model or deterioration; and
ii. vehicle operating and road maintenance cost models.

Table 5.5 Impact of operating conditions on truck life.

Typical depreciation hours	Increasing application severity	Typical severity indicators
15,000	Low severity	Earthmoving and stockpile use with well-matched loading equipment
		Free flowing material
		No impact loads
		Short to medium hauls
		Well-maintained level roads
10,000		Road-building, dam construction, open-pit mining
		Varying load conditions
		Some impact loads
		Varying haul road conditions
		High rolling resistance and poor traction in parts of haul cycle
		Some adverse grades
8,000		Poorly matched loading equipment
		Many impact loads
		Continuous overloading
		Continuous use on poorly maintained haul roads
		High rolling resistance and poor traction throughout
	High severity	Frequent adverse grades in haul cycle

Source: Caterpillar, 2017.

The approach is suited to a network of mine haul road segments, as opposed to a single road analysis. For a number of road segments of differing functional and traffic volume characteristics, together with user-specified road maintenance and VOC unit costs, the model assesses the following:

- Traffic volumes applied to the road network segments over the analysis period (as specified)
- The change in road condition (assessed as rolling resistance), by modelling the effects of truck speed, traffic volume and wearing course (as a function of the wearing course parameters controlling rolling resistance)
- The maintenance quantities as required by the particular strategy
- The vehicle operating costs (by modelling)
- Total costs and quantities
- The optimal maintenance frequency for specified segments of the network such that total costs are minimised

Figure 5.23 illustrates the MMS flow chart used to determine the optimum maintenance interval for a mine road network consisting a number of variable road segments. Cost savings associated with the adoption of a maintenance management system approach are dependent on the particular hauling operation, vehicle types, road geometry and tonnages hauled, etc. The model can accommodate various combinations of traffic volumes, road segment

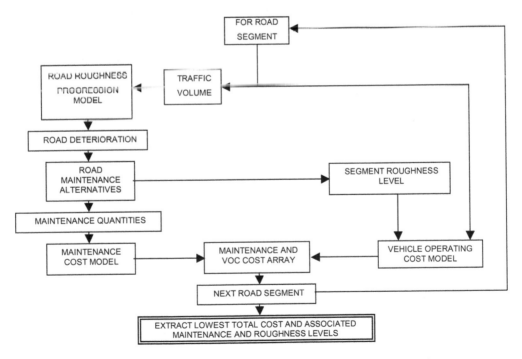

Figure 5.23 Flow diagram of MMS for mine haul road network (for a single maintenance strategy iteration).

geometries and wearing course material properties to enable a full road network simulation to be completed.

The first element of a MMS for mine haul roads is based on modelling the variation of vehicle operating costs with rolling resistance. When combined with a road maintenance cost model, the optimal maintenance strategy for a specific mine haul road, commensurate with lowest overall vehicle and road maintenance costs may be identified.

5.3.7 Example of MMS application

The objective of producing a MMS model for a mine haul road network is to evaluate alternative maintenance strategies within a system of constraints related to total cost and maintenance quantities such that the optimal maintenance policy for the network, commensurate with lowest total costs, can be identified. The basis of the evaluation is total cost, consisting of haul truck fuel, tyre, parts and labour costs together with road maintenance costs incurred when using both the motor-grader and water-car. Road construction and vehicle depreciation costs were not considered since these will be similar irrespective of the maintenance strategy evaluated. Truck operator costs, although significant, vary widely across sites and operations and likewise are not considered here.

Initially, the mine haul road network is divided into a number of specific road segments of homogeneous characteristics corresponding to changes in haul road geometry (width, grade), wearing course material types and the daily tonnage a particular segment carries. Table 5.6 shows the basic data required for a network of five roads, HR-01 to HR-05. Data generic to the network of roads is also defined at this point, as per Table 5.7.

The model is then developed by following the process depicted in Figure 5.23 by determining the following:

- The increase in rolling resistance, using the equations presented in Section 4.4.2, associated with an increasing maintenance interval
- The truck speed and fuel burn, at each maintenance interval, following the Cooper equations referred to in Section 5.3.1.2 and 5.3.2
- Tyre, parts and labour costs at each maintenance interval, following Section 5.3.4 and 5.3.5
- Road maintenance costs incurred for maintenance interval, following Section 5.3.3.

Figure 5.24 shows a summary of the various cost and performance results by following the above procedure for the road network segments and data described in Tables 5.6 and 5.7.

Table 5.6 MMS road segment data.

Road Segments Specific Data	HR-01	HR-02	HR-03	HR-04	HR-05
Road length (m)	2160	1300	1800	1200	950
Width (m)	35	35	35	35	40
Grade (%, uphill +ve)	10	8	0	3	9
Vehicle speed estimate (km/h)	10	15	50	25	15
Daily tonnage (kt)	43	61	30	77	66
Shrinkage product	177	174	213	90	354
Grading coefficient	18	7	15	15	7
Plasticity index	4	4	11	4	17
CBR (%) 100% Mod AASHTO	66	78	38	38	59

Table 5.7 MMS Generic data.

Truck Data

Engine power (kw)	1566
Transmission efficiency (%)	80
Retarder efficiency (%)	115
Unloaded weight EVM (t)	324
Loaded weight GVM (t)	143
Speed limit (laden/unladen) (km/hr)	57
Replacement cost ($\times 10^6$)	5.00
Average fleet engine hrs ($\times 10^3$)	10.00
Truck utilisation (% per day)	70
Tyre costs (per tyre)	45,000
Fuel cost (per litre)	1.26
Labour cost per hour	50.00
Labour factor β	0.45

Grader Data

Grader effective blade pass width ($L_e - L_o$)(m)	3.00
Number of graders	2
Road m²/h per m effective blade pass width	6000
Grader utilisation (% per day)	70
Grader operating costs/h	250

Water Cart Data

Water cart effective spray width (m)	12.00
Number of water carts	2
Road m²/h per m spray pass	12,500
Water cart utilisation (% per day)	70
Water cart operating costs/h	300

The resulting MMS for this example is shown in Figure 5.25 from which it is seen that each road segment has a characteristic cost profile and optimum maintenance interval at which costs for that segment are minimised. Optimum maintenance interventions are thus:

- HR-01 = every 5th day
- HR-02 = every 6th day
- HR-03 = every 3rd day
- HR-04 = every 4th day
- HR-05 = every 2nd day

It should also be noted that each segment also has a cost penalty associated with too-frequent road maintenance, where the cost of road maintenance is not recouped through reduced VOC. Using this approach, it is also important to check whether or not the available road maintenance equipment can meet the schedule described above. If not, consideration should be given as to which roads should enjoy road maintenance priority, in terms of the cost change associated with an increase in the road maintenance interval beyond the optimum indicated. Thus, in Figure 5.25, the 'post-maintenance' gradient of each cost profile would dictate which road segments enjoy priority; HR-05 having the greatest increase per day should be prioritised, whilst in contrast segment HR-02 shows the lowest increase per day and road maintenance could be deferred on this segment to focus resources on segments where maintenance generates greater benefits.

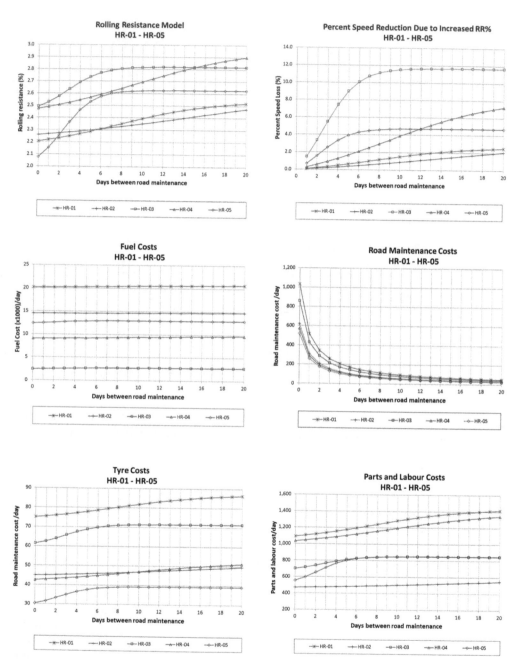

Figure 5.24 MMS cost summaries for example application.

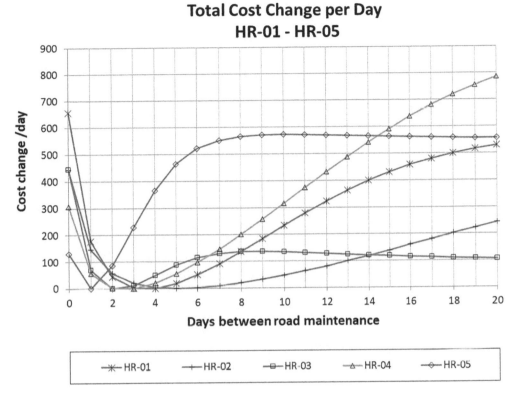

Figure 5.25 Variation of total cost changes per day across network of roads HR-01 – HR-05

Finally, using segment HR-04 in the earlier example, Figure 5.12 is reconstructed in Figure 5.26 to show how road maintenance and VOCs combine through an MMS to deliver the optimal maintenance interval and associated total costs.

Dunston *et al.* (2007) and Lee (2010) further developed the MMS approach described here to consider the effect of reduced production rates associated with road maintenance activities, arising from the use of either alternative routes or an alternate lane of the same haul road. Once road maintenance is completed, the haul road is restored to its required level of functionality (and rolling resistance) condition and the next 'unrestricted' hauling cycle (nc) starts. Models were proposed to track the influence of both rolling resistance and the variables of cost in the total operation cost for a given hauling project. The total cost for a hauling operation (TC) is equal to the product of hourly hauling costs (HC_{haul}) and the sum of the unrestricted operation time (t_{uop}) and the maintenance time (t_{mtce}), plus the sum of the hourly maintenance cost (HC_{mtce}) multiplied by the maintenance time (t_{mtce}), for each operation-maintenance cycle. The hourly hauling costs refer to ownership cost, fuel cost, labour cost, tyre cost, etc., whilst the hourly road maintenance costs consist of the equipment (e.g., motorgrader and water truck) costs. Equation 5.17 describes the model proposed by Dunston *et al.* (2007):

$$TC = \sum_{cycle=1}^{nc} \left(HC_{haul} \cdot \left(t_{uop} + t_{mtce} \right) \right)_{cycle} + \sum_{cycle=1}^{nc-1} \left(\left(HC_{mtce} \cdot t_{mtce} \right) \right)_{cycle} \qquad \text{Equation 5.17}$$

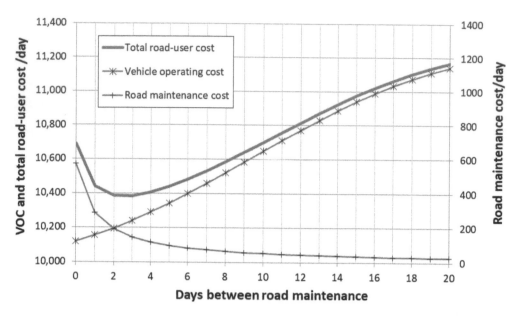

Figure 5.26 Using HR-04 example data to determine the optimal frequency of road maintenance commensurate with minimum total costs.

With this approach, the critical assumptions for the general case are described as follows:

* Maintenance equipment is being charged only for actual maintenance periods.
* Haulage and road maintenance hourly costs are constant for the project.

The authors thus ignore the increase in required rimpull (due to greater rolling resistance), which leads to an increase in the real hourly cost of haulage operation, since a single
hourly cost is typically assumed for the equipment over the duration of an entire project.
Conversely, HC_{mtce} is often charged for the full duration of the project regardless of actual
utilisation, but the previous model assumes maintenance is charged only for the actual maintenance hours used. That is, the focus is on cost rather than pricing.

Lee (2010) used this model as a basis for investigation after how much change of rolling
resistance should maintenance be triggered, and how rapidly is a maintenance trigger level
reached. To achieve this, he considered the economic efficiency of using mine equipment as
a platform for real-time rolling resistance monitoring, based on earlier work and conceptual
development by Thompson *et al.* (2003, 2006a) and Miller *et al.* (2004).

5.4 Real-time road maintenance

The most cost-effective approach to mine road network maintenance management is based
on a real-time system which integrates truck- and pavement-interaction data as a basis for
making road management-based decisions. Most large surface mines already utilise a high
precision GPS and data communications system backbone for asset management purposes.

In addition to managing truck allocation and routing throughout the mine, they also interrogate various original equipment manufacturers' (OEM) truck on-board vital signs monitoring systems. Typical of such systems are the Cummins Cense engine condition monitoring system, the Komatsu vehicle health monitoring system (VHMS), Komatsu Komtrax or Caterpillar vehicle information management system (VIMS) which monitor a range of vital vehicle operational parameters, as switched, analogue, pulse width modulated, frequency or calculated values, depending on machine type (Caterpillar, 1999). Selected VIMS data also forms the basis of a qualitative road analysis control system (RACS) (Caterpillar, 2000) in which strut pressure, speed and mode of operation data are extracted to determine vehicle racking (twisting or pothole effect), pitching (bounce or ditch/hump effect) and load bias (road crown effects). However, whilst the RACS system provides a useful qualitative and benchmarkable assessment of application severity, it does not have the ability to recognise specific road defects nor rolling resistance.

Using a typical mine-wide communication and truck GPS location system as a starting point, Thompson *et al.* (2003) proposed a real-time maintenance management system (RT-MMS) (Fig. 5.27), showing the initial developmental links and how existing communication, location and truck monitoring systems are integrated and the information from the system applied in making road maintenance decisions. From a haul road test consisting of 81 typical mine road defects over an area of 24,000 m², including potholes, fixed stones, corrugations, humps and ditches of various 'degrees' of defect, a total of 189 triggers were recorded for 16 truck repetitions over the course. Using the selected accelerometer trigger sensitivity

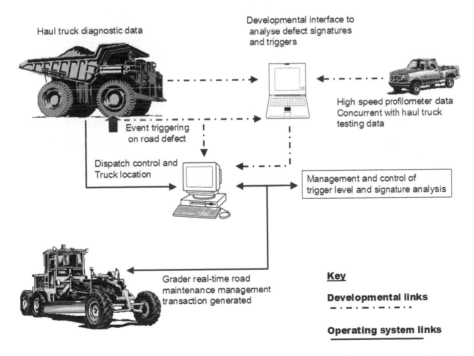

Figure 5.27 Real-time mine road maintenance system concept and integration with existing mine-wide communication, location and truck monitoring systems.

Source: Thompson *et al.*, 2003.

of ±0.6g, 88% of the defects were detected during fast unladen runs, the number reducing with reduced truck speed and under laden conditions. For all the runs combined, 7% other extraneous triggers were recorded. By combining on-board truck diagnostic information with this data, it was possible to determine the source of these additional defect-unrelated triggers. In the majority of cases, these were ascribed to harsh braking, gear change (torque converter lock-up to first gear) or centrifugal effects during sharp, high speed manoeuvres.

Further work (Hugo *et al.*, 2008) and mathematical modelling developed a more rigorous analytical methodology to integrate multi-sensor information to eventually isolate, recognise and dimension a road defect from its vibration signature and other on-board diagnostic data sources. The approach adopted was based on the development of a multi-body dynamics model of the truck under consideration, used to solve the so-called inverse problem that entails finding the road input from measured responses and known system characteristics.

This approach requires a fairly difficult characterisation exercise to determine the specific haul truck system geometry, spring stiffness coefficients, damping coefficients, as well as mass and inertia characteristics. A number of these characteristics have to be experimentally determined, both from field and laboratory testing. In some cases fairly simple experiments are possible in which, for instance, the truck tyre deflection characteristics are measured. Other components such as the suspension struts have to be characterised using servo-hydraulic actuators and special jigging to identify the non-linear stiffness and damping characteristics of these struts. The chief advantage of this characteristically intense approach is system versatility in the sense that it is possible to computationally correct for varying loads, speeds and other parameters that change continuously. This may also enable the recognition system to be universally applied across a wide range of truck types and sizes.

Using this approach, Hugo *et al.* (2008) demonstrated that the simple single-degree-of-freedom model shown in Figure 5.28 of the unsprung mass on an independent front suspension system provided an adequate description of the vehicle response dynamics. One of the important features of this simplified model was that neither elastic structural vibration, nor vibrations induced by other sources on board (such as the engine) significantly contaminate the measured signals in the frequency band below 10 Hz. It is this frequency range that is both typical and characteristic of road-vehicle interactions. Assuming the acceleration of the unsprung mass of the truck z''_u and the force acting between the sprung and unsprung masses F_{susp} to be measured as functions of time, the force exerted on the unsprung mass m_u by the tyre F_{tyre} can be determined from the equation in Figure 5.28.

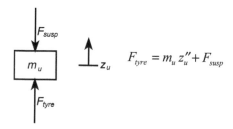

Figure 5.28 Simple single-degree-of-freedom model of truck independent front unsprung mass.

Once the tyre force is known, the unsprung mass displacement and an inverse tyre model are required to calculate the corresponding road input. Treating the tyre as a simple linear spring with stiffness k_{tyre} allows inversion through Equation 5.18:

$$Z_r = z_u + \frac{F_{tyre}}{k_{tyre}}$$

Equation 5.18

where Z_r is the road elevation (a measure of the defect dimensions, m), and z_u is the unsprung mass displacement (m). The unsprung mass displacement is derived by double integration of the unsprung mass acceleration. Using the known speed of the vehicle, the resulting profile was transformed to the spatial domain to reconstruct the road defect geometry, examples of which are shown in Figure 5.29. However, some practical limitations exist when attempting to reconstruct road defects, especially where truck speed is high and or defect density on the road is high, coupled with gear change, braking and rate-of-turn induced dynamic events. However, the defect location and density map concept provide a useful indication of which active haul roads would benefit from a maintenance intervention. When the system is extended to the truck fleet as a whole, then the resulting defect density and maintenance priority or 'heat' map will enable a mine to evaluate road conditions on a real-time basis, with resultant improvement in service and reduction in total road-user costs.

Proof Engineers (2017) have developed the RT-MMS approach further to include a low-cost 'bolt-on' road condition monitoring (RCM) system to record and transmit a road condition score for site benchmarking and road maintenance management purposes. Following

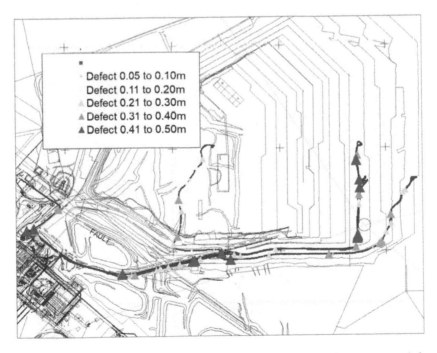

Figure 5.29 Real-time road defect geometry measurements. Symbols represent defect magnitude (depth or height).

calibration to site and vehicle, it delivers a road condition score (Fig. 5.30) based on the International Roughness Index (IRI), which in real time gives a quantifiable and objective assessment of road and pit conditions. Although it simplifies to some extent the haul truck characterisation exercise of Hugo's approach, it nevertheless suffers from some of the same issues with respect to separating driver-and truck-induced responses from true road responses.

The road condition monitoring system provides full site mapping, showing the state of the road in a simple colour graded format indicating the different levels of road condition or defect severity, allowing required maintenance locations to be identified and road maintenance triggered. A similar approach has been reported by Brown *et al.* (2002) in which Canadian forest roads were evaluated through the use of truck vibration signatures (near real-time but without further vehicle on-board data input with which to qualify the vibration records).

Other approaches to RT-MMS include Joseph *et al.*'s (2017) field evaluation of rolling resistance using each tyre of a mine truck as a measurement device, capturing the tyre load response and coupling this with a cyclic plate load test of various locations on a haul road to establish the resilient pressure stiffness and its variation along a road. The latter supplies the ground deformation, which when evaluated in parallel with the tyre deformation establishes the tyre-ground contact area and thus permits the rolling resistance to be evaluated. As with

Figure 5.30 Example of road condition monitor (RCM) output.
Source: image courtesy of Proof Engineers.

other RT-MMS systems, mapping the rolling resistance magnitudes by GPS truck location as a truck moves around the mine site identifies the critical zones requiring maintenance attention on a real-time basis.

Thompson (2017) applied the 'live' speed-rimpull-gradeability spline estimation approach (Fig. 5.20) to predict road rolling resistance, based on real-time speed, engine-throttle position and power data from on-board systems, coupled with high-precision road elevation and GVM data. Although some variance in rolling resistance was evident from point to point, as shown in Figure 5.31, over a longer haul at consistent speeds (such as ascending a ramp or a flat-haul), it was possible to detect changes in rolling resistance in terms in a shift of the average values recorded over a period of time. However, as with other rolling resistance–based RT-MMS systems, the approach has a number of limitations, especially relating to data precision and truck dynamics, such as braking, deceleration and cornering, etc.

The key to all the RT-MMS systems is both the measure of road condition and how that condition is evaluated in terms of its impact on total road user costs. Whilst rolling resistance is the most widely understood parameter from a mine operator's perspective, it is especially problematic to measure on a real-time basis and indirect measurements may prove to be the most tractable approach in the future. However, since most large mines operate ultra-heavy mine trucks with on-board diagnostic data collation, relaying data within a broader asset management and centralised communication, cloud storage and GPS backbone, it is

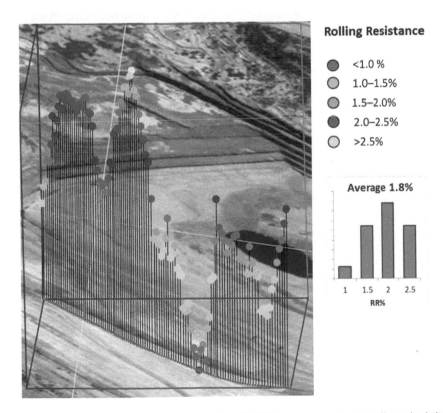

Figure 5.31 Estimation of rolling resistance for RT-MMS using speed-rimpull gradeability and on-board data, plotted along selected portion of laden truck haul.

recognised that road condition can be monitored on a real-time basis and the data integrated within the centralised truck management system.

Once road condition information (either as defects, IRI or rolling resistance, etc.) is received by the communications hub on the mine and the appropriate weighting applied (for traffic volumes, condition severity, etc.) a transaction can then be initiated to automatically inform road maintenance assets of the location, type and, critically, priority. Road rolling resistance can also be deduced on a similar basis, thus areas of high rolling resistance (which may not be associated with a particular road defect, but rather a high density of defects on a given section of road) can also be recognised and repaired on a real-time basis. By combining on-board data and defect recognition within the broader mine-wide asset management and communication system, other associated benefits derived from this approach are as follows:

- Event map histories which show consistently poor sections of road requiring betterment or rehabilitation
- More effective utilisation of existing road maintenance assets: rapid response to identi-fied road defects
- Reduced capital expenditure on road maintenance assets: expanding road network can be effectively maintained with less equipment
- Immediate recognition of haul road conditions – visual inspection for each change of hauler route unnecessary
- The increased utilisation of the existing computer-based mine and transport management system to provide stream-lined and integrated data management and information

Chapter 6

Equipment performance and costing road construction

6.1 Introduction to accounting analysis

There are two specific operations in the accounting analysis, namely an economic analysis to justify projects and project alternatives, and financial analysis to determine the budget and cost of a project. In this context project is defined as a haul road project. Extensive feasibility studies are performed on the viability of new mines, including cash flow and return on investment. There are handbooks that focus on these types of analyses and project, which are beyond the scope of this book.

In an economic analysis of road projects the question is which alternative is the most economical? For example, a shorter steeper route or a longer flatter route? Or the economic viability of say importing a dump rock material or using a weaker in-situ material as a base. In such analyses a life-cycle cost is determined, and it is not a precise exercise as principles or design standards are required. This type of analysis is discussed in the next section.

In many instances a construction budget is required to allocate funds to be able to execute the project. These calculations have to be precise, as incorrect budgeting could have serious implications. Input for this type of analysis is given in Section 6.3.

6.2 Economic analysis

6.2.1 Background

The practice of engineering involves many choices among alternative designs, procedures, plans and methods. Since the available alternative courses of action involve different amounts of investment and different prospective receipts and disbursements, the question "Will it pay?" is nearly always present. This question may be broken down into subsidiary questions:

Why do this at all? Should a proposed new activity be undertaken? Should an existing activity be expanded, contracted or abandoned? Should existing standards or operating procedures be modified?

Why do it now? Should we build now with excess capacity in advance of demand, or with only sufficient capacity to satisfy the immediate prospective demand? Are the costs of capital and other business conditions favourable to a present development?

Why do it this way? The choice among alternative ways of doing the same thing is common to all types of engineering activity.

The central problem is how to judge whether any proposed course of action will prove to be economical in the long run, as compared to other possible alternatives. This is performed by an economic study, which may be defined as a comparison between alternatives in which the differences between the alternatives are expressed so far as practicable in monetary terms.

6.2.2 Project evaluation

The most widely used technique for economic evaluation of alternatives is cost-benefit analysis. Although the analysis is termed "cost-benefit", a number of measures such as net present value (NPV), internal rate of return (IRR) and benefit-cost ratio (B/C) are used. The procedure involves a number of steps:

Identification of alternatives – All possible alternatives which would achieve the set goals need to be identified. These should not be limited to alternatives proposed by proponents, but the net should be cast wider and other, possibly innovative, alternatives sought. One of the alternatives should be the null alternative – that is, the present situation – or the do-minimum alternative. It will serve as a basis for the measurement of marginal costs and benefits.

Identification and quantification of benefits and costs – The alternatives to be considered then need to be explicitly defined. Quantification requires that relevant proposals, plans, designs and data be assembled.

The benefits and costs of each alternative then need to be quantified. The costs include taxes for private entities, whereas road projects for government would exclude taxes as the government is the recipient of taxes and it would result in double counting. The benefits of an alternative could include lower costs to the user, improved safety, and reduced time. The economic costs of an alternative are the value of resources lost to the economy because they are used in the execution of the project. They will normally include the costs of materials, labour, services and capital goods.

In the case of projects having long life spans, and where uncertainties regarding the future could impact on the economic viability of the project, it may be advisable to repeat the analysis or do a sensitivity analysis for different future scenarios.

Assessment of alternatives – The alternatives are then assessed using economic principles. Several different assessment techniques may be used, all of which require the same data. They will all select the same alternative as the best alternative, provided the correct procedures are followed. The following are some of the project assessment criteria that may be used:

Discounted values: All costs and benefits are expressed in terms of the value at the beginning of the analysis period. Future costs are converted by the discount rate, which is the opportunity cost of capital. This value is not related to the inflation rate or prime interest rate, but it reflects the potential cost of borrowing. Typically, a financially strong company would use a lower discount rate than a troubled company. In the event of uncertainty, a range of discount rates should be used to evaluate project alternatives.

If $b_0, b_1, \ldots, b_n$ are the benefits of a project, typically the reduced transportation costs

        $c_0, c_1, \ldots, c_n$ are the costs of a project, typically construction and maintenance costs

in years 0, 1, . . ., n

and   i is the discount rate

then present value of the benefits is $b_0/(1 + i)^0 + b_1/(1 + i)^1 + \ldots + b_n/(1 + i)^n$

and   present value of the costs is $c_0/(1 + i)^0 + c_1/(1 + i)^1 + \ldots + c_n/(1 + i)^n$

Net present value: The difference between the benefits and costs (the net benefit) in the specified year is discounted to the present by using the discount rate. The discounted sum of all these net benefits over the economic project life is defined as the net present value (NPV), according Equation 6.1:

$$NPV = \sum\nolimits_1^n \frac{b_n}{(1+i)^n} - \sum\nolimits_1^n \frac{c_n}{(1+i)^n}$$ Equation 6.1

A positive result indicates that there is a net benefit, and that the project is economically justified. A negative result indicates a net disbenefit if the project were to be executed. The criterion for the acceptance of a project is that the net present value must be positive; in other words, funds will be voted for a project only if the analysis produces a positive net present value. Where a choice has to be made between mutually exclusive projects, the project with the largest net present value will be chosen since it maximises the net benefit to the mine.

Internal rate of return: The internal rate of return (IRR) is the discount rate at which the present values of costs and benefits are equal. It is therefore the value of the discount rate (r) which satisfies the Equation 6.2:

$$\sum_1^n \frac{b_n}{(1+r)^n} - \sum_1^n \frac{c_n}{(1+r)^n} = 0$$ Equation 6.2

Only projects with an internal rate of return higher than the discount rate, which forms a lower limit, will be considered for funding. If a number of alternatives are considered, the alternative yielding the highest internal rate of return is not necessarily the preferred one.

Discounted benefit-cost ratio: The discounted benefit-cost ratio (BCR) is the ratio of the present value of benefits relative to the present value of costs:

$$BCR = \frac{\sum_1^n \dfrac{b_n}{(1+i)^n}}{\sum_1^n \dfrac{c_n}{(1+i)^n}}$$ Equation 6.3

A project will be considered for funding only if the benefit-cost ratio is greater than 1. Again, if a number of alternatives are considered, the alternative yielding the highest benefit-cost ratio is not necessarily the preferred one. To determine this, an incremental analysis should be undertaken.

Incremental analysis: The previous assessment criteria utilise benefits and costs relative to the null alternative. In an incremental analysis, benefits and costs relative to other alternatives are used. The alternative with the lowest initial capital cost, and with an acceptable

IRR (greater than the discount rate) or BCR (greater than 1), is taken as the first defender. All other projects with acceptable IRR or BCR are taken in order of increasing capital cost and in turn become attackers. An attacker defeats the present defender if the IRR or BCR computed using the additional benefits and additional costs is acceptable. Should an attacker defeat the defender, that alternative becomes the new defender. The preferred project is the final defender.

6.2.3 Discussion of project analyses

Project analyses are applied to long-term facilities such as the main surface or principal ramp road, which have a life in excess of 10 years. Similarly such analyses could be conducted to define mine policy with respect to say the use of a specified base. Conversely, there is little point in analysing a bench road with an anticipated life of weeks or months. There is little argument about the need for such a facility, as this is part of the mine plan.

In an economic analysis inflation rate is not considered, as it is assumed that all the costs and benefits would be subject to a similar inflation rate, and alternatives would be similarly affected. Additionally it is pure guesswork to predict the long-term, say 10 years, inflation rate.

Most mine haul roads have no further use once the mining activity is completed, and thus the residual value is ignored. If the road has to be removed, there may be a cost implication for restoration. However, the discounted cost would be relatively small and would typically have little influence on the outcome between alternatives.

6.3 Equipment performance and costing for road construction

In the mine planning phase, the provision of reliable and realistic costs for equipment (capital and operational) is critical to the economic evaluation of a mining project or even to determine the economics of individual tasks within the operation. In the case of large volume, high capital, low unit excavating or transporting cost equipment, the reliability and impact of these assessments becomes critical in the overall economic model and resulting mine plan. This section is not intended to be a guide to costing complete in itself but rather to demonstrate some of the typical approaches that can be used and introduces some of the basic terms, concepts and techniques involved.

Haul road construction will take time and incur costs for a mine. Interestingly, poorly performing roads take time and money to be built as also roads that perform well, but with time it becomes clear that good haul roads will out-perform the poor roads and consequently the maintenance costs will also be lower. Design and construction costs for the majority of haul roads represent only a small proportion of the total (life of road) operating and road maintenance costs. This is especially true in high-tonnage (and often low-grade) mining where several haul roads form the 'backbone' of the operation over many years. The performance of these haul roads will affect several other aspects regarding the mining costs that are not only related to the construction costs. For example, rolling resistance will directly affect the operational cost by increasing the cycle times and energy (diesel) consumption; maintenance costs will be impacted when a road rapidly deteriorates, or requires frequent rehabilitation (replacing the wearing course due to degradation or erosion), tyre consumption increases when road surface conditions deteriorate and operator health and ergonomics (often a hidden cost that is difficult to measure unless associated with vehicle speed reduction or multiple driver change-overs in-shift), amongst others. All these examples indicate in the general sense that good roads will provide a reduction of these cost-impacts and consequently in unit costs.

As a primary cost consideration, first assess the in-situ conditions to determine the appropriate project design strategy considering the geometry, structural and functional aspects according the category of each section of the road, as described in the previous chapters. The category of road is broadly divided into three classes: permanent, semi-permanent and temporary, considering the level of usage and permanence during the mine life. As an example, assume that a permanent road will be constructed and define general aspects that will be incorporated as construction costs:

• Subgrade preparation, including cut and fill to achieve the desired geometry
• Subbase transportation, placement and preparation if necessary
• Base material transportation, placement and preparation
• Wearing course material production (when needed), transportation, placement and preparation
• Berm placement and construction
• Lateral drainage
• Peripheral drainage
• Traffic management (signage), etc.

The construction material is typically and ideally comprised of materials found in the mine lease area. However, as is often the case with new mines in areas of deep weathering or regolith, the absence of suitable materials might require the acquisition of the appropriate material from external areas or even purchased from a third party. The latter two options are clearly more expensive due both to the haulage distance involved and the sheer volumes required for a typical mine road. Using the structural design presented in Chapter 3 for a typical 226 t payload truck together with the road width determinations from Chapter 2, a 33 m pavement construction width is required (29 m roadway and 2 m shoulders/side drain each side). This would require approximately 8.5 t wearing course (at 200 mm thickness), and 27 t selected blasted hard rock for the base layer (at 370 mm thickness) per linear metre of haul road.

The ideal situation therefore would be to source the appropriate material from on-site sources (mainly and most cheaply waste or overburden materials) and that could be brought to the road construction site directly by large mine trucks, either as part of the production process, or (less ideally) being excavated and loaded from a suitable mine dump (re-handling of the material will increase unit costs of construction). As the waste material would need to be moved from the mine to waste dumps anyway, diverting some material directly to construction would represent a saving in waste transportation and consequently could be disregarded from the transportation cost when building the road. Another relevant consideration (especially for low stripping ratio operations) is that waste dumps are generally limited in size due to environmental considerations, so any material that is sent to road construction (or left in the pit) would reduce required waste dump volumes slightly, representing a saving for the mining project.

The previous comments would apply equally to the wearing course material, where this is sourced directly from waste stripping. However, in most cases, suitable material for a wearing course is sourced from a small surface crushing operation (which also often provides blast stemming as a 'by-product') and in these instances double handing is unavoidable. Although a cost for construction, the value in preparing a wearing course that meets the specifications discussed in Chapter 4 have been clearly illustrated in Chapter 5. Typical surface crusher installations are shown below in Figure 6.1.

Construction of the roadside safety berms is often made without due regard to the source of the material and how that material may interact with the wearing course itself. Ideally, berms should be constructed from material similar to the wearing course to prevent contamination of the road surface, which otherwise could cause rapid deterioration of the road. Another important consideration for the berms is the ability to shape them to a steep inner face to better redirect a wayward vehicle and also to ensure that the berms themselves contain no material which could damage truck tyres. Therefore free-dump material is not ideal for these safety features and consideration should be given to using wearing course materials for these too, or at least the berms constructed with waste rock should be dressed with crushed wearing course material.

Figure 6.1 Examples of surface crushing and sizing installations for preparation of mine haul road wearing course material.

6.4 Earthworks equipment and typical operations

As an introduction to cost-estimation for haul road construction, the approach used here is to firstly determine the main types of equipment used and then the number of operating hours required from each item of equipment to complete its specified construction task. Approaches for the hourly basis are given in AusIMM (2012). Although it does not discuss the construction detail of a haul road, it is a good reference to estimate operating hours for the major mining equipment. Tannant and Regensburg (2001) also give a simple calculation to estimate the costs of construction of a haul road.

Primary earthworks (layerworks) are one of the most important elements in road construction because it establishes a stable foundation to provide the structural and functional integrity to the haul road. A roadway on a substandard foundation will fail prematurely, which means that the road's layerworks are as important as the finished surface. The main tasks during a mine road construction project are typically the following:

- If required, provide land clearance (above ground), grubbing (roots and stumps, etc., below ground level) and top/sub-soil removal, for later ground surface rehabilitation.
- If necessary, establish new local drainage pattern in the vicinity of construction.
- Establish vertical alignment by creating cuts and embankment fills (ideally from cut material). Work embankments (fills) in the down-grade direction, and place cuts working in the up-grade direction to control heavy rainfall.
- Shape and compact the subgrade to required specification. During this stage, the geometric design must be established with the crown or camber applied to assist in drainage during construction into the lateral drains in its final design. Drainage is a critical element because improper drainage will hamper construction and reduce layerworks strength, thereby reducing new road performance and life expectancy.
- Following subbase or cut and fill works, the remaining layerworks are placed according to the specific design structural design and methodology adopted. In the mechanistic structural design methodology, most often a selected hard, rocky material is used as previously discussed. A large (tracked) bulldozer is well suited to perform this task, together with end-tipping by the construction trucks, in lifts (thicknesses) dictated by the compaction equipment used. Each lift must be well compacted by suitable equipment, as the dozer and truck pass combinations do not provide the compactive effect required to achieve material interlock and thus layer strength. Note that when a selected blasted rocky material is used, no water is required during compaction.
- To complete the earthworks, the wearing course material is placed according to the functional design specification. This layer is spot-tipped in the centre of the road, opened and then moistened to achieve OMC, following which the layer is shaped and compacted in one or more lifts depending on the design thickness.

Table 6.1 shows the corresponding equipment necessary to build the road and the stage that each piece of equipment would be employed during the road construction.

The number of hours each piece of equipment spends to perform a task will depend on the number of units assigned to the task, the volume or area to be worked, the specific task each equipment is executing, the type of material that it is being handled and the level of finishing and category of road envisaged in the design. With regards to volumes of material, it

Table 6.1 Main equipment used in road construction at each stage.

	Platform	Berms	Drainage	Sub-grade	Sub-base	Base	Wearing course
Compactor	☑			☑	☑	☑	☑
Dozer	☑	☑	☑	☑	☑	☑	☑
Excavator/Loader	☑	☑			☑	☑	☑
Motor grader	☑	☑	☑	☑			☑
Dump truck	☑	☑			☑	☑	☑
Water truck	☑						☑

is important to consider the state of the material and how it changes throughout the dig, haul and dump/compact processes. Three states are commonly used:

*   Bank state: A material (clay, sand, gravel, rock, etc.) that has not been disturbed from its natural state.
*   Loose state: A material that has been disturbed (loosen) by excavating, working or stockpiling.
*   Compacted state: A material that has been loosened, spread and then worked by compaction.

Typical conversion factors are given in Table 6.2 relative to bank volumes (BCM – bank cubic metres), for LCM – loose cubic metres and CCM – compacted cubic metres. These are also referred to as bulking factors. Individual sites and materials would have to be evaluated to determine local variations on these generic values. The density associated with these volumes would, especially for soils, require consideration of moisture content to determine the actual (soil) density, for estimating purposes.

6.4.1 Typical excavator/loader applications

An excavator or loader, depending on the characteristics of the material handled (in situ, level of fragmentation, stockpiled) will be necessary to load or reload the material taken from the mine or disposed in an intermediate stockpile containing the appropriate material that will be used to build a specific layer. The number of hours spent by the equipment used in this task is dependent on the cycle time that comprises the number of passes considering the tasks listed here:

*   Excavating/loading to fill the bucket
*   Swing/transportation loaded
*   Dumping in the truck
*   Swing/displacement empty

6.4.2 Typical truck applications

Trucks will be necessary to move the materials from the source of material to the road site under construction. It can be considered in most circumstances that the material used to build

Table 6.2 Typical conversion factors between BCM, LCM and CCM.

Material	From\To	BCM	LCM	CCM
Clays	BCM		1.43	0.90
	LCM	0.70		0.63
	CCM	1.11	1.58	
Gravels and Sands	BCM		1.15	0.96
	LCM	0.89		0.87
	CCM	1.04	1.17	
Rock (Blasted)	BCM		1.40	1.25
	LCM	0.71		0.86
	CCM	0.80	1.16	

the haul road will be formed essentially by run-of-mine waste (sourced from the mine). But production mining equipment may not be designated for this service due to the impact it may have on the mining cycle and production. If the material taken is run-of-mine, normally the (production) truck will dump the material as close as possible in a stockpile and then another excavator or loader will reload in a (generally smaller) construction truck for dumping on the construction itself, for subsequent shaping, mixing or opening, etc., by other equipment (dozer or grader for example). This may not be the case if the dump rock layer is constructed, as it may be more productive to eliminate double handling. So essentially the number of construction truck hours assigned to the task of road construction will be based on the cycle times that trucks will need to perform:

- Spot and load at the source of material
- Travel laden from the source or stockpile of material to the road site
- Spot and dump the material at the road site
- Return empty to the source of material

As in all truck and excavator operations, the truck size should be matched to the size of the loader/excavator used, according to the typical match rule which considers the truck must be filled in three to seven bucket loads from the loader/excavator employed.

6.4.3 Typical tractor/dozer applications

A bulldozer is a crawler (tracked tractor) equipped with a substantial metal plate (known as a blade) used to push large quantities of unconsolidated or lightly compacted soil, sand, blasted rock or other such unbound material during construction and earthworks. It is typically equipped at the rear with a claw-like device (known as a ripper) to loosen densely compacted materials. The equipment is suited for heavy tasks and can be used to handle coarse fragmented material (typically used in the base layer), cutting and filling the in-situ material to prepare and level the subgrade or even opening the material used to build the wearing course. It is a versatile piece of equipment and also central to any haul road construction project. Applicable sizes are generally in the >50 t GVM, 300 kw engine power class. Many larger bulldozers are equipped with electronic position controls that can be programmed to construct the geometric design.

Bulldozer production can be estimated using the production curves (Fig. 6.2) together with appropriate correction factorising, following Equation 6.4:

$$P_D = CP.PFC$$

Equation 6.4

Where

P_D = Production (measured in loose cubic meters) (Lm³/h)
CP = Estimated production (from Fig. 6.2) (Lm³/h)
PFC = Production factor corrections (maximum three corrections applied)

Production (loose cubic meter per hour) can be estimated for dozers units larger than 300 kW engine size and 4.5 m (universal) blade width from Figure 6.2.

$$CP = 4608e^{\left(\frac{0.0212E_P}{Bw}\right)}\left(d_d^{-0.875}\right)$$

Equation 6.5

Where

E_P = Engine power (flywheel) (kW)
Bw = Universal blade width (m)
d_d = Dozing distance (m)

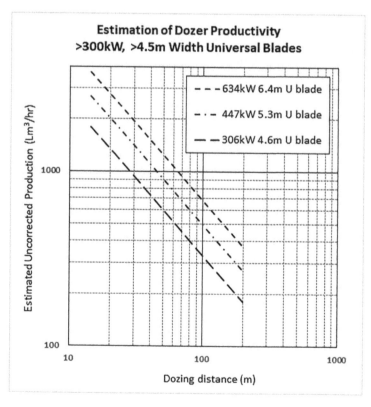

Figure 6.2 Dozer productivity chart (for equipment with >300 kW power and >4.5 m U-blade).

The production curves estimate maximum (uncorrected) production for universal blades and are based on the following conditions:

- 60 minutes per hour operating
- Flat dozing grade
- Power shift machines with hydraulic blade controls
- 15 m cut (gear 1F) then drifts (2F) blade load to dump over edge, return in gear 2R.
- Soil density of 1370 kg/Lm^3
- Coefficient of traction >0.5

Correction factors (PFC) are based primarily on:

- Operator skill, from 0.6 to 1.0 (novice to highly experienced)
- Material type, from 0.6 to 0.8 (rock – ripped or blasted) to 1.2 (loose stockpile)
- Visibility – reduce by factor of 0.8 for compromised visibility (darkness, rain, dust etc.)
- Blade – adjust production by ratio of selected blade SAE capacity compared to Universal blade.
- Grade of doze, from 1.2 at 10% downgrade to 0.8 at 10% upgrade

6.4.4 Typical motor grader applications

A motor grader, also commonly referred to as a grader, is a construction machine with a long blade used to level and create a flat surface during the grading process. Typical models have three axles, with the engine and cab situated above the rear axles at one end of the vehicle and a third axle at the front end of the vehicle, with the blade in between. Capacities range from a blade width of 2.5 to 7.3 m and engines from 93 to 373 kW.

In earthworks, the grader's primary purpose is to 'finish grade' (to refine or set precisely), although some 'heavy-blading' may also be undertaken. Where 'rough levelling' is required this is performed by heavy equipment such as dozers. In the construction of haul roads graders are used to prepare the base layer to create a wide profiled surface upon which to place the wearing course. Graders can produce cambered or crowned surfaces, together with the drainage ditches (with shallow V-shaped cross-sections) on either side of the haul road. Details on the type of activities a grader performs in road construction are given below, whilst grader productivity estimation has been discussed in Chapter 5.

6.4.4.1 Site preparation

This application involves any light material cutting, moving, and mixing necessary to prepare a site for road construction. A variety of materials are encountered in this application and blade loads vary depending on the activity being performed. Both heavy blading and finish blading are performed when preparing a site and pass lengths are typically in the range of 30 to 300 m. The production of a grader is given essentially by the type of service and the operating speed that the equipment will develop during the task execution and the blade's effective length considering the angle of positioning. Typical operating speeds for site preparation vary depending on whether heavy blading or finish blading activities are being performed but are generally under 15 km/h (refer Figure 5.9).

6.4.4.2 Heavy blading

This application involves cutting, moving, and mixing material, usually in the initial stages of surface preparation. A variety of material types are moved in this manner and the blade tip position varies accordingly. Full blade loads are usually experienced during heavy blading, since moving material is the primary goal. Pass lengths within this application vary, but are usually less than 600 m. Unlike finish blading, the speed of the machine is dependent on the load being moved and typical operating speeds are under 10 km/h, therefore, gears 2 through 4 are generally used in this application.

6.4.4.3 Finish grading

This application involves preparing a haul road surface finish and would also include the maintenance of mine haul roads, as discussed in Chapter 5. During construction, the material being moved is usually a moist, dense wearing course – but only relatively light blade loads are encountered after the initial open and spread of the wearing course is complete. Following initial compaction, there is a 'final-cut' performed which provides a smooth consistent running surface. With 'finish' or 'final-cut' blading, the motor grader operator skill level is the highest and requires the highest degree of accuracy. Thus, it is primarily done at low operating speeds – usually less than 5 km/h – in gears 1 and 2 and to ensure a smooth, even finished surface, the same gear is usually maintained for a given pass. Pass lengths during this application are usually less than 600 m for road construction and 150 m for site development.

6.4.4.4 Ripping/Scarifying

This application involves conditioning hard surfaces before they are bladed, or preparing a surface to receive a new wearing course cover so as to prevent laminations and improve bonding between the two layers of material. Shanks on the ripper and/or scarifier are pushed into the ground, thus breaking up otherwise hard surfaces, which makes grading operations less damaging to the mouldboard. Rippers and scarifiers can also be used to mix aggregates together, although over longer distances, a more efficient mix is achieved from an offset disc towed plough. Rippers generally penetrate 150 to 300 mm into the ground, while scarifiers typically penetrate to a depth of 25 to 200 mm. Pass lengths are generally less than 600 m for both activities and the typical maximum speed for this application is about 6 km/h in gears 1 or 2. If the ripper/scarifier is used for mixing aggregates, the typical operating range becomes 6 to 20 km/h in gears 3 to 6.

6.4.5 Typical water truck applications

A water truck is used during road construction to provide the necessary moisture to the layers under construction to guarantee an effective compaction of each layer (apart from when a selected blasted waste rock base layer is used where water is not required). The unit can additionally be used for general dust suppression during construction and if fitted with a water-cannon, is often deployed to water down dumps or stockpiles when loading (when not deployed for its primary purpose). The water truck hours of operation can easily be determined considering the speed of watering, the number of passes required to bring the material to OMC (i.e. the material moisture content at the start of compaction) and the spray delivery rate and pattern width as discussed in Chapter 2, together with the construction length and width. Depending on the evaporation rates, multiple applications may be necessary. Typical watering speed during construction is 18 to 30 km/h, generally in second or third gear. The volume of water required per lift per meter length of road can be determined from Equation 6.6:

$$VW = \frac{\rho_{dry}\left(OMC - \omega_f\right)Rw\, h}{100}$$ Equation 6.6

Where
VW   = Volume of water to be added (or evaporated) per metre length of road (m³)
ρ_{dry} = Material dry density (kg/m³)
OMC = Optimum compaction moisture content (practically slightly dry of optimum) %
ω_f = Moisture content in field (borrow pit) (%)
Rw   = Compacted width of road (m)
h   = Compacted layer thickness (m)

Similarly, the rate (l/m²) is found from Equation 6.7:

$$Rate = \frac{\rho_{dry}\left(OMC - \omega_f\right)h}{100}$$ Equation 6.7

6.4.6 Typical compactor applications

To achieve the specified layerworks compaction at the lowest cost, the correct selection of compaction equipment is critical. Selection is dependent on the type of material, its compactability and the mine site conditions. A wide range of compaction equipment may be used, depending on site conditions and materials being compacted. For pavements, some kind of wheeled or vibrating roller is often used. The suitability of the various equipment options is summarised in Table 6.3.

On mine haul roads, the layerworks compaction process may be continuous and as such the compactor's size and productivity should complement the hauling and spreading equipment used. Manoeuvrability is less of a factor as compactors on mine roads will frequently shuttle back and forth in order to reduce turnaround time. Wide roads and thicker lifts or selected waste rock layers will require the largest rollers available in order to maximise productivity.

The productivity of a compactor is given by a relationship which includes the width of the roller, the speed of displacement and the weight of effective compaction depth. This productivity determines the thickness of compaction that is typically 100 to 200 mm and considering the area (or volume) of the site to be compacted. As commented in Chapter 5, an impact roller can be used to produce more effective compaction and productivity in mining haul road construction considering it has a deeper effect allowing thicker layers (up to 1 m in thickness) to be compacted. With any equipment, passes should be limited to between 4 and 20. If the required passes are excessive, alternative equipment for compaction should be investigated.

Table 6.3 Compaction equipment application summary.

Equipment	Typical operating speed (km/h)	Best results	Good results	Poor result
Smooth wheeled vibrating rollers	6	Well-graded sand-gravel, crushed rock, asphalt	Running surface, base courses, subgrades	Uniform sands, Clays, silty clays
Rubber tyred (pneumatic) rollers (proof rollers)	12	Coarse grained soils with some fines	Running surface	Coarse uniform soils and rocks
Grid rollers	20	Weathered rock, well-graded coarse soils	Subgrade, subbase	Clays, silty clays, uniform materials
Sheeps-foot rollers, static	6	Fine-grained soils with > 20% fines	Dams, embankments, subgrades	Coarse soils, soils with cobbles, stones
Sheeps-foot rollers, vibratory	5	Fine-grained soils with > 20% fines, but also sand-gravel mixes	Subgrade layers in similar soils	
Impact rollers	15	Most saturated and moist soils, gravels and sands, sub-grade subbase and base, dump rock layers	Dry, sands and gravels, sub-base and base	

Compaction production can be estimated from Equation 6.8 as:

$$P_{compact} = \frac{16\,W_r\,v\,h\,ProdF}{BF\,n_{passes}}$$

Equation 6.8

Where

$P_{compact}$ = Production (measured in loose cubic meters) (Lm³/h)
W_r = Width of roller or rolled path (m)
v = Speed (km/h) (In vibrating equipment, frequency and working speed should be adjusted to one impact per approx. 25 mm travel. When speed is too high a 'washboard' type finish results, but each compaction application will be different, and a good operator will adjust the balance of frequency and speed to most efficiently compact the layerworks material).
h = Compacted layer thickness (m)
$ProdF$ = Productivity factor (allows for time losses associated with turns, operator skill, etc). Varies from 0.9 (excellent, long compaction or shuttle runs) to 0.7 (poor, short compaction runs with turnarounds)
BF = Material bulking factor
n_{passes} = Number of passes required to achieve compaction specification

From which the number of similar units can be determined once the total LCM of the layerworks is known.

6.5 Earthworks equipment capital and operational costs

Capital and operating costs are fundamental to any economic analysis of equipment, mine planning or feasibility studies, budgeting and data for analysis of different mining sub-system options. These costs are, however, often difficult to estimate reliably. Sources of data are typically these:

- The company's own operations
- Operating mines with a history of using similar equipment configuration
- Previous studies
- Equipment manufacturers or suppliers
- Consultants, government and industry authorities
- Contractor quotations
- Rules and formulae. Textbooks and handbooks have some formulae but these are often too generalised and are more geared to local markets. For these to be used successfully, the local cost factors are required – and these are not simply an exchange rate correction, but must also consider relative cost of parts, labours, fuel, etc.

When attempting to use or corroborate costs from several sources listed previously, there are often inconsistencies which make the selection of a representative cost figure somewhat qualitative. These discrepancies arise due to variable working condition factors, quality of maintenance, location, mine accounting practices, or abnormally high or low costs at a particular mine which are not fully reported or benchmarked across the industry. This results in the necessity for using a systematic approach to cost estimation. The advantages and importance of determining costs from first principles and then cross-checking them are these:

- Costs can be developed for any mining equipment, not just equipment already in use on the mine site.
- There is consistency between costs and estimation assumptions used and this consistency can be checked, or the assumptions modified, to generate a rough indication of cost sensitivity or assumption risks.
- A cost data base is the starting point for an economic analysis of a mine. A significant component of which is the equipment selection and costing aspect.

With regards to the costs associated with earthworks equipment the costs are mainly divided into capital costs (CAPEX) and operational costs (OPEX). Capital costs are fixed, one-time expenses incurred on the purchase of land, buildings, construction, and equipment used in the production of goods or in the rendering of services. In other words, it is the total cost needed to bring a project to a commercially operable status. On the other hand, operating (operational) costs are the expenses which are related to the operation itself, or to the operation of a device, component, piece of equipment or facility. They are the cost of resources used by an organisation just to maintain its existence.

6.5.1 Estimating cost of operation

The recommended approach is to build up operating cost estimates from first principles, then cross check the costs thus derived with actual mine statistics, global cost databases, cost-estimation handbooks etc. and whatever 'rules of thumb' are available. The approach has been generalised to enable the basic principles to be applied to any piece of mining equipment used in road construction. The steps involved in deriving operating costs are as follows:

* Split the machine into defined cost elements, such as

    * power or fuel
    * lubrication and filters
    * tyres or crawler and undercarriage
    * repair parts and major overhauls
    * wear parts

* Subdivide each cost element into component parts
* Assign a life or utilisation to each component part and calculate the hourly cost
* Total all the components to achieve the total hourly operating cost

Deriving costs from first principles is certainly the most reliable method. However, determining the costs and life for all components in a machine is quite time consuming and it is often difficult to get accurate data or data that is not skewed by site-specific conditions. To overcome this problem, a set of factors or formulae can be used to describe, in general terms, the conditions under which a piece of equipment is operating. Overall working conditions can be classified into the three main categories described earlier, but attention should be given to the critical role of the following:

* Material characteristics such as density, bulking, abrasiveness, hardness
* Water and dust
* Working conditions – especially operator skills, climatic conditions and altitude
* Labour factors including management, maintenance philosophy, skill-sets and availability of spare parts
* Utilisation factors such as the annual operating hours, average engine load factors on the final estimate of operating costs

6.5.1.1 Fuel or power costs

Fuel consumption can be easily measured in the field. However, if no opportunity exists to do this, consumption can be predicted when the machine application is known. Consumption is based on engine size (kW), consumption in litres per kW and load factor. Several simulation packages can also determine fuel consumption, based on engine-torque and fuel consumption maps. Similarly, equipment suppliers offer fuel consumption estimates based on job application load factors. The load factor controls engine fuel consumption, which varies with periods spent at idle, drive, manoeuvring at part throttle and predominantly downhill operations, all of which impact the load factors.

6.5.1.2 Lubrication and filter costs

If no detailed lubrication costs are available, they can be calculated as a percentage of the fuel cost. For most non-hydraulic machines (or only minor hydraulic components), 15% is typical, which may increase to 30–40% for equipment with significant hydraulic system (such as a hydraulic excavator). However, these larger systems are also more sensitive to severe duty, heavy dust, deep mud or water and can increase this proportion by up to 25%.

6.5.1.3 Tyres or crawler and undercarriage

Tyre costs are obtained by multiplying the cost of each tyre by the number of tyres and dividing by the life in hours. Tyre manufacturers give guidelines for calculating life, recommendations and maintenance procedures, based on a series of reduction factors for base average life (Goodyear, 2009, 2017). Base average tyre life is adjusted according to specific conditions to obtain approximate estimated hours (or kilometres) as the final product. Condition factors include the following:

- Maintenance quality
- Operating speed (maximum)
- Surface conditions
- Wheel positions (especially driven)
- Loading
- Curves/Manoeuvring
- Grades (driven only)

For tracked machines, the hourly cost of the undercarriage is often calculated separately as a wear item rather than being included in the repair and maintenance cost breakdown. This is because there is poor correlation between undercarriage environment and application. Consider, for instance, on a rock fill application, which is an extremely abrasive, high-wear environment, but if it is generally a flat down-hill short push, the machine itself works in an essentially easy application. The three conditions that affect life-expectancy of track-type undercarriages to the greatest extent are these:

- Impact
- Abrasiveness
- Environmental, operational and maintenance

6.5.1.4 Repair parts and overhauls

The single largest item in operating costs are generally repair parts. Hourly operating costs are adjusted upwards as the unit ages and averaging of these costs tends to over-estimate costs for newer units whilst under-estimating them for older equipment. The cost of repair parts is difficult to estimate and in any specific application, actual cost experience on similar work provides the best basis for establishing the hourly repair costs. Equipment applications, operating conditions and maintenance management would mostly influence relative repair costs.

Where no historical data is available, a method based on the assumption that any piece of equipment is a collection of spare parts could be used. Although part life varies, assuming a generic operating life, it is possible to calculate the total costs of the major repair parts expected to be purchased over this life, and therefore the hourly cost.

Major overhaul costs can be estimated as a percentage of initial capital cost or as a build-up of components and their life. For example, a truck could be subdivided into engine, transmission, body, frame, electrical and so on. The cost of each of these major components (or the cost of rebuilding them) can then be estimated with the estimated life. This gives a standard cost per hour even though the actual expenditure may only occur when the damage or rebuild is implemented. A typical estimate of overhaul costs is 15% of capital cost every 10,000 hours.

6.5.1.5 Wear parts

Wear parts are often referred to as operating supplies or ground engaging tools. In construction, they include high-wear items such as bucket teeth, ripper boots, cutting edges, wear plates and liners, etc. These are usually separately itemised as they are directly related to the ground conditions. An approximate method is to take a factor of the capital cost which is the same logic such as for maintenance supplies.

6.5.2 Estimating cost of ownership

To protect equipment investment and be able to replace or recoup equipment value, the mine must recover over the machine's useful life an amount equal to the loss in resale value plus the other costs of owning the equipment. For most construction equipment, there is no resale value at end-of-life and the original equipment investment is recovered by establishing depreciation schedules according to the various uses of the equipment. Ownership costs can be established either by calculating the sum of the straight-line depreciation and an additional percentage to cover other ownership cost items, or by considering the equipment as if it were being leased.

6.6 Cost comparison of road design options

In the previous chapters, various technical approaches to road design and management were discussed. However, in many cases the 'value proposition' only examined the potential savings (or conversely, the cost penalties) associated with road improvement strategies. This value would be generated through haulage unit cost reductions arising from increased vehicle speeds, reduced cycle times, fuel burn, tyre wear, etc. Added incentives were highlighted when the cost and frequency of road maintenance is considered. However, there needs to be an implementation cost, associated with the road-building or repair/rehabilitation which would initially offset some of the value these improvements create.

In this section, a cost model is illustrated as a basis for costing and comparing the various road-building and repair or rehabilitation options previously discussed. The comparison made here is based on the cost of hours of activities, which means the number of hours spent by each equipment and the final costs calculated according cost estimation handbooks (Info-Mine, 2017; Western Mine, 2002). The costs provided by these manuals and references are typical costs taken from the tables that are divided in capital costs and hourly operating costs. The operating costs are mainly divided into the following items:

- Overhauls for parts and labour
- Maintenance (parts and labour)
- Fuel (depending on the energy used by each piece of equipment, e.g. diesel, gasoline, electric power, natural gas)
- Lubricants

- Tyres
- Wear parts

The sum of those items is considered the total hourly operating costs for each equipment, which, multiplied by the number of hours spent in each activity would define the total operating costs to execute such activity.

6.6.1 Cost estimation example

To illustrate the approach to cost estimation to build a road, the example provided in Section 3.3.4 is used where a comparison between CBR and the mechanistic structural designs approaches is made when using a CAT793D truck. Considering a stretch of road of 1000 m length and construction width of four times the width of the truck (with an operating truck width of 8.3 m) results in 33.2 m roadway plus shoulders and a space to build the lateral safety berm and drains. The material to build all layers is taken from the sources and/or stockpiles that are assumed to be stored 1500 m away from the construction site. The key truck specifications used in the construction and necessary for the calculations are presented in Table 6.4.

By way of example, the difference in materials used and primarily the difference in layer thickness (as presented in Table 6.5) will deliver different unit costs of construction for the each design options.

Starting with the CBR-based design approach example, the compacted volume of the subbase is:

$$V_L = R_L R_W h$$

Equation 6.9

Where
V_L = Compacted volume of the layer (m³)
R_L = Road Length (m)
Rw = Road width (m)
h = Layer Thickness (m)

Table 6.4 Key specifications of design truck.

CAT 793D Key Specifications.	
Operating Width (m)	8.30
Payload (t)	229
Volume (m³) Heaped 2:1	176
Gross Vehicle Mass (t)	384

Source: Caterpillar, 2015.

Table 6.5 Pavement design layer thicknesses determined for the CBR- and mechanistically-based structural designs.

	CBR	Mechanistic
Wearing course	200 mm	200 mm
Base	300 mm	500 mm
Sub-base	300 mm	
Subgrade	Semi-infinite	Semi-infinite

Using the numerical example:

$$V_L = 1000 \times 33.2 \times 0.3 = 9960 \, m^3$$

Equation 6.10

The same logic is used for the other layers of the two designs, resulting in the volumes to be used in the road construction as presented in Table 6.6.

The equipment used at each stage were presented in Table 6.1 and the estimation of the number of hours spent by each equipment during construction is based on cycle times, speeds and average performance of the equipment used during the execution of the different tasks. For example, to build the sub-base the following equipment is used:

i.   the excavator to dig/load the material;
ii.   the trucks to transport from the loading area to dump in the construction site;
iii. the dozer to blade, push and spread the material; and
iv.   a roller to compact the material.

The operating cost estimations used were sourced according to INFOMINE Cost Estimation 2017 (InfoMine, 2017) for each item of equipment considered in this example. These figures can be updated using new cost estimation guides, or even better, with the owner's operating costs at a specific mine site. To calculate the operating costs (OPEX), each cost extracted from the cost estimation tables was then multiplied by the operating hours calculated for each piece of equipment.

6.6.1.1 Excavator hours

It is assumed that to dig, excavate, swing and dump, the cycle time takes 0.5 minutes for each pass, a swing factor of 1 for a swing angle of 90^0 and a bucket factor 1 (fill factor/swell factor), the number of hours of this equipment is calculated accordingly:

$$N_L = \frac{L_v}{B_s}$$

Equation 6.11

Where
N_L = Number of loads
L_v = Layer volume (m³)
B_s = Bucket size (m³)

Table 6.6 Compacted volumes necessary for each road layer to build the 1 km stretch for the CBR- and mechanistically-based structural design options.

	CBR Design	Mechanistic Design
Wearing course (m³)	6640	6640
Base (m³)	9660	16,600
Sub-base (m³)	9960	0

$$N_{eh} = \frac{N_L \, C_{et}}{60}$$ Equation 6.12

N_{eh} = Number of excavator hours
C_{et} = Cycle time of the excavator for each load in minutes
Table 6.7 summarises the excavator operating hours.

6.6.1.2 Truck hours

The cycle time of the truck is highly dependent on the average haulage distance and of the speeds that the equipment is able to develop during the transportation from the loading point (source of the material) to the dumping point (construction site). For this example, the assumed speeds and times are shown in Table 6.8.

The number of truck loads is calculated by adding the partial times of the entire truck cycle as shown in Equation 6.13, Equation 6.14 and Equation 6.15:

$$T_L = \frac{HD}{V_F} \, 60 = \frac{1.5}{12} \, 60 = 7.5 \, min$$ Equation 6.13

Where
T_L = Time loaded (min)
HD = Distance from source to construction site (km)
V_F = Speed loaded (km/h)

Table 6.7 Excavator operating hours required to build the two design options.

	CBR					Mechanistic				
	Thickness (mm)	Number of loads	Cycle time (min)	Operating hours (h)	OPEX	Thickness (mm)	Number of loads	Cycle time (min)	Operating hours (h)	OPEX
Wearing course	200	1186	0.5	9.88		200	1186	0.5	9.88	
Base	300	1779	0.5	14.82		500	2964	0.5	24.70	
Sub base	300	1779	0.5	14.82		0	–	0.5	–	
Total (h)				39.52	4512				34.58	3948

Table 6.8 Speed and time for the truck cycle.

Speed loaded (km/h)	12
Speed unloaded (km/h)	30
Spotting and loading time (min)	2
Spotting and unloading time (min)	2

$$T_E = \frac{HD}{V_E} 60 = \frac{1.5}{30} \times 60 = 3.0\,min$$ Equation 6.14

Where
T_E = Cycle time to return empty (min)
V_E = Speed for the empty truck (km/h)

$$T_T = T_L + T_E + T_{SL} + T_{SU} = 15.5\,min$$ Equation 6.15

Where
T_T = Total truck cycle time (min)
T_{SL} = Spotting and loading time (min)
T_{SU} = Spotting and unloading time (min)

The number of loads to bring the material from the source to the construction site is given by the following:

$$N_{TL} = \frac{V_L}{V_{Truck}}$$ Equation 6.16

Where
N_{TL} = Number of truck loads
V_L = Compacted volume of each layer (m³)
V_{TRUCK} = Volume capacity of the truck (m³)

For the sub-base layer construction using a 50 m³ capacity truck, consider the CBR design:

$$N_{TL} = \frac{9960}{50} = 199.2\,loads$$ Equation 6.17

The number of truck hours (N_{TH}) for this layer is then

$$N_{TH} = N_{TL}\,T_T = \frac{(199.2 \times 15.5)}{60} = 48.14\,h$$ Equation 6.18

Repeating the earlier approach for layers in both design options gives the results in Table 6.9.

6.6.1.3 Dozer hours

The productivity of a dozer is estimated using the charts presented in Figure 6.2. Assuming a dozing distance of 60 m for a Caterpillar D9 equivalent using a semi-universal blade would produce 1000 m³/h, consequently the number of hours would be:

$$N_{DH} = \frac{V_L}{P_D}$$ Equation 6.19

Where
N_{DH} = Number of dozer hours
P_D = Dozer productivity

The results are given in Table 6.10.

Table 6.9 Truck operating hours required to build the two road designs.

	CBR					Mechanistic				
	Thickness (mm)	Number of truck loads	Truck Cycle time (min)	Operating hours (h)	OPEX	Thickness (mm)	Number of truck loads	Truck Cycle time (min)	Operating hours (h)	OPEX
Wearing course	200	217	11.00	39.78		200	217	11.00	39.78	
Base	300	325	11.00	59.67		500	542	11.00	99.46	
Sub base	300	325	11.00	59.67		0	–	11.00	–	
Total (h)				159.13	11,155				139.24	9761

Table 6.10 Dozer operating hours required to build the two road designs.

	CBR					Mechanistic			
	Thickness (mm)	Dozer production (m³/h)	Operating hours (h)	OPEX	Thickness (mm)	Dozer production (m³/h)	Operating hours (h)	OPEX
Wearing course	200	1000	6.64		200	1000	6.64	
Base	300	1000	9.96		500	1000	16.60	
Sub base	300	1000	9.96		0	1000	–	
Total (h)			26.56	3417			23.24	2989

6.6.1.4 Motor grader hours

The motor grader productivity is estimated using the concept of the effective blading width and overlapping width between passes (according to section 5.3.4 in Chapter 5), the type of work the equipment is doing (heavy blading, finishing, ditching) and consequently the speeds that the equipment is going to develop during each of these tasks. Speed for the different motor grader tasks ranges typically between 3 and 10 km/h (according to Figure 5.9).

$$N_{GH} = \frac{Area}{V_G \left(Ef_{BL} - Sup_{Lat} \right)} 1000 \, Job_{ef} \qquad \text{Equation 6.20}$$

Where
N_{GH} = Number of operating hours of the motor grader
Area = Surface area of the road (Road length × Road width)
V_G = Average grader speed during the execution of the task (km/h)
L_e = Effective blade length (m)
Sup_{Lat} = Lateral superposition between passes (m)
Job_{ef} = Working efficiency considers a number of factors that can impact in the result of the grading task and are related to working conditions, operator skills, etc.; typical efficiency factors vary from 0.7 to 0.85.
Table 6.11 summarises the operating hours associated with both design options.

Table 6.11 Motor grader operating hours required to build the two road designs.

	CBR					Mechanistic				
	Thickness (mm)	Heavy Blading Hours	Finish Blading Hours	Ditching Hours	OPEX	Thickness (mm)	Heavy Blading Hours	Finish Blading Hours	Ditching Hours	OPEX
Wearing course	200	5.67	14.17	9.45	1931.46	200	5.67	14.17	9.45	1690.02
Base	300	8.50	21.26	14.17	4828.64	500	14.17	35.43	23.62	4225.06
Sub base	300	8.50	21.26	14.17	3219.09	0	–	–	–	2816.71
Total (h)		**22.67**	**56.68**	**37.79**	**9979.19**		**19.84**	**49.60**	**33.06**	**8731.79**

6.6.1.5 Compactor hours

The compactor is considered to be a vibratory roller doing five passes for every 200 mm of each layer plus the in situ material for site preparation. Considering the width of the roller, the width of the road to be built and speed of the roller during operation the compactor hours are calculated as shown in Table 6.12.

6.6.1.6 Water truck hours

The estimation of the water truck hours during construction (Table 6.13) is considered to be equal to the number of hours calculated for the roller, as the water truck is required to apply water to the material during compaction. Wetting is not necessary for the construction of the dump rock base, considering the size of the particles that constitute this layer.

6.6.2 Evaluation of design options costs and benefits

The cost estimation comparison of the two designs introduced previously is made by summing the operating costs (OPEX) from the different operations involved in building each design. The construction cost per unit square metre can then be found, together with the percentage contribution of each activity and layer to the total costs as shown in Table 6.14 .

The difference in costs between the two projects is obvious considering the increase in thickness and consequently in volume required for the CBR design. The mechanistic design technique is the more cost effective solution, and the CBR design method would result in over-expenditure on a design that would in all likelihood not perform to expectations.

Next, the influence of building a substandard road structure is considered. The consequence of reduced structure and construction cost would impact the quality of the road by requiring a higher frequency of maintenance and the expected rolling resistance would also be higher. Considering that an increase of 1% in rolling resistance typically represents a 10% decrease in speed on the ramp roads (and 22% on the level roads – the specific loss of speed can be determined from the Equations presented in Chapter 5). Operational cost included is the value of losses to be analysed from the financial point of view to justify the investment in building a haul road according specification.

Consider a road that is at a 10% grade plus 2% rolling resistance, which represents a high standard road, as reflected by the mechanistic design with a 500 mm dump rock base.

Table 6.12 Roller operating hours required to build the two road designs.

	CBR				Mechanistic		
	Thickness (mm)	Compacting Hours	OPEX		Thickness (mm)	Compacting Hours	OPEX
Wearing course	200	14.34			200	14.34	
Base	300	21.50			500	35.84	
Sub base	300	21.50			0	–	
Total (h)		**57.34**	**2798.05**			**50.17**	**2448.30**

Table 6.13 Water truck operating hours to build the two road designs.

	CBR				Mechanistic		
	Thickness (mm)	Watering Hours	OPEX		Thickness (mm)	Watering Hours	OPEX
Wearing course	200	14.34			200	14.34	
Base	300	21.50			500	–	
Sub base	300	21.50			0	–	
Total (h)		**57.34**	**2798.05**			**14.34**	**699.51**

Table 6.14 Haul road OPEX distribution considering the various equipment necessary to build the road.

	Cost Distribution			
	Design 1 (CBR)		Design 2 (Mechanistic)	
Truck	11,156	29.15%	9761	29.15%
Excavator	4512	11.79%	3948	11.79%
Tractor	3417	8.93%	2990	8.93%
Grader	9979	26.08%	8732	26.08%
Compactor	2798	7.31%	2448	7.31%
Water truck	6405	16.74%	5604	16.74%
Total	38,267	100.00%	33,484	100.00%
Unit cost (US$/m²)	**1.15**		**1.01**	

The costs as calculated in Table 6.14 for construction is US$1.01/m². For the design of longitudinal grade plus rolling resistance (effective grade of 12%) the average speed for the CAT 793D standard truck (laden) is 11 km/h (Caterpillar, 2015). Considering the costs of construction presented earlier, comparative costs are calculated for seven different designs by incrementing the dump rock base thicknesses by 50 mm (with the wearing course a constant 200 mm thickness throughout) as given in Table 6.15.

For example, if the mine decides to build a road with 300 mm thickness of base (instead of the 500 mm recommended design provided by the mechanistic approach) to save costs and time during construction, there are consequences. Assuming that the lower performance of the road translates to an increase in rolling resistance from 2% to 3%, the impact in the associated loss of truck speed on operating costs are given in Table 6.16. The second column represents the increase in costs considering a progressive decrease in speed (considered 10% reduction at each 1% increase in rolling resistance).

Considering as a reference the OPEX of the 227 t truck which according to INFOMINE (2017) is US$ 292.09/h, it is assumed that this cost is valid for a 2% rolling resistance road as shown in Table 6.16. For a load of 227 t for 1 km of road the cost is 0.13 US$/t, with an increase in rolling resistance to 3% the reduction in speed results in an operating cost increase to 0.14 US$/t. Furthermore, assume that this operation produces 2 Mtpa and that all mine production passes over this road.

As an investment consider the two designs (mechanistic) and the 'do minimum design' with insufficient thicknesses of base and an out of specification wearing course (200 mm thickness throughout) used to build the road. Considering the construction costs obtained from the equipment performance and effective working hours shown in the previous tables, the same calculation can be followed to determine the costs of construction for a number of non-standard alternatives, as presented in Table 6.17.

Considering that the mechanistic design, plus the selection of the appropriate material for the wearing course, outperforms the design provided by the 'do minimum' approach and

Table 6.15 Unit construction costs for different thicknesses of dump rock base in different designs.

Base Structural Design Thickness (mm)	Cost (US$/m²)
200	0.58
250	0.65
300	0.72
350	0.79
400	0.86
450	0.94
500	1.01

Table 6.16 Increase in truck operating cost considering the increase in rolling resistance.

Rolling Resistance (%)	Op cost (US$/trip)	Op. Cost (US$/t)	Percent increase
6%	48.68	0.21	66.67%
5%	41.73	0.18	42.86%
4%	36.51	0.16	25.00%
3%	32.45	0.14	11.11%
2%	29.21	0.13	0.00%

Table 6.17 Costs of construction of a 1 km road for different dump rock base thickness in the structural design.

Base Structural Design Thickness (mm)	Cost (US$/m²)	Total Cost (US$)	Comments
200	0.58	19,133.65	
250	0.65	21,525.35	
300	0.72	23,917.06	Do minimum design
350	0.79	26,308.76	
400	0.86	28,700.47	
450	0.94	31,092.17	
500	1.01	33,483.88	Mechanistic design

Table 6.18 Difference in operating cost considering the increase in rolling resistance.

Rolling Resistance	Op. Cost (US$/t)
3%	0.130
2%	0.117
Difference	0.013

the reflection of that affects the frequency of maintenance of each of those designs and also the rolling resistance. This is the premise for the analysis of the investment made in the construction of a haul road according the standards given throughout this book.

Accept the 'do minimum' as the null hypothesis (or base case) of investment in the construction of the design, defined as H_0.

Accept the different designs as the alternative to the 'do minimum' case defined as H_1. The difference between $H_1 - H_0$ will be considered as the investment (I_0). The investment is given by the difference between the two designs from Table 6.17.

Investment H_0	US$ 23,917
Investment H_1	US$ 33,483
Difference in Investment (I_0)	US$ -9,566

The benefits of the investment will be given by the improvements from H_0 to H_1 in terms of rolling resistance and consequently expressed in terms of savings in operating costs, denoted as B_n. The reduction in costs (referred here as benefits) is given by the difference between cost of 2% RR and 3% RR from Table 6.16 and multiplied by the production during the period of analysis (2 Mtpa), resulting in US$25,994.

Considering that the improvements are going to be analysed distributed during one year of production, this benefit (2 Mtpa × 0.013 US$/t = US$25,994) has to be divided into 12 months ($b_0, \ldots, b_n$, according to Fig. 6.3), resulting in US$2,166.23/month and the value will be brought to present value at a monthly discount rate of 1.2%. The NPV of this improvement is a positive value (Equation 6.21) demonstrating the viability of the construction of a project according to the standards in the expectation of keeping the rolling resistance at 2%

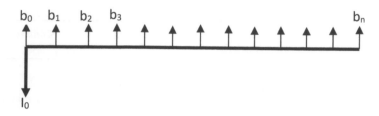

Figure 6.3 Investment cost (I_o) and benefit (b_n) analysis of haul road construction following specifications.

and consequently developing the speeds on the road according to those provided by the truck performance curves.

$$NPV = \sum_{1}^{n} \frac{b_n}{(1+i)^n} - I_0 = \sum_{n=1}^{12} \frac{2166.23}{(1+0.012)^n} - 9566 = \text{US\$ } 14{,}510 \qquad \text{Equation 6.21}$$

The internal rate of return (IRR) of the comparative analysis can also be calculated, and gives a result of 32.6%, demonstrating that the investment in the construction of a haul road according to specifications is a highly recommended project for investment. The calculation and the improvements shown here are only considered for a 1 km length of a road. If these returns are multiplied by the haul road network of a mine, or if assuming that the rolling resistance difference will be even higher, or considered over a longer period of time, it emphasises the philosophy of designing a haul road network according to the procedures described throughout the book. Besides the improved operational aspects, there is a significant opportunity to increase profits.

Chapter 7

New technology and haulage equipment developments

7.1 Background and orientation

The mining industry is constrained by decreasing resource grades, energy and labour cost increases, stringent safety and environmental controls and capped capital and working-cost considerations. However, it is also looking into a future of increasing global demand for many of its products. It could be argued that the outlook is not dissimilar to that of the previous century, it is only the level of technology that was appropriate then, as compared with now, that guides the solution strategy. Open-pit mining began in earnest in the early 1900s and led indirectly to the first dump truck patent in 1920 by Mawhinney, following which truck capacity has grown to upwards of 360 t today. This was accompanied by a reduction in mineral grades mined, implicating larger volumes of material to be moved at increased levels of efficiency, together with lower overall margins on the 'value' each truck load would generate.

Associated with this mining method development are various technological initiatives, from early open-pits in the 1900s, the block caving methods of the 1950s, programmable logic controllers in 1970s, through to automation, autonomous vehicles, autonomous haulage systems (AHS) and, ultimately, the autonomous or 'Mine of the Future™' as envisaged by Rio Tinto and others (Rio Tinto, 2018). From the start of the 21st century, mining has evolved from the idea of a 'modern' mine, to that of a 'real-time' mine and, ultimately, will evolve into an 'intelligent' mine, as discussed by Pukkila and Sarkka (2000). Figure 7.1 shows this evolution and the accompanying development of autonomy, from simple user-interface and monitoring development, with minimal data and analytics, through to more process and analytically complex and data-rich aspects of perception, position, navigation and mission planning and independent equipment collaborative processes.

Automation, robotics and access to rich data and the supporting analytical processes are key themes throughout the industry. Big data analytics – the process of examining large and varied data sets to uncover 'hidden' data, establishes new correlations and performance measures that can help operations to optimise, 'monetise' the value of the data and use the insights they deliver to improve operating efficiencies and methods. Figure 7.2 shows how data volume and analysis have evolved towards predictive models supported by 'big data' analytics and Artificial Intelligence as opposed to the essentially reactive processes and reporting that still predominates to a large extent today.

For mining, although the vision of a 'future mine' varies between commodities, mining methods and individual operations, they all envisage a mine based on increased automation, connectivity and greater application of Information and Communications Technology.

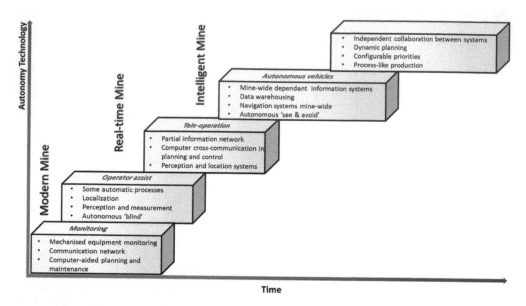

Figure 7.1 Incremental evolutionary phases of autonomous haulage and mining systems.
Source: modified after Pukkila and Sarkka, 2000.

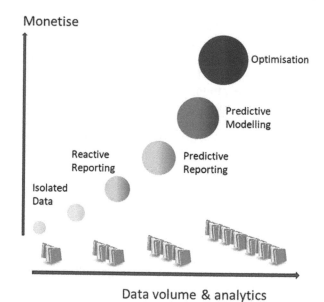

Figure 7.2 Opportunities to monetise the value of big data through optimisation of mining processes.

Connectivity includes people, processes, information and technology. These insights into the future of mining commonly point to several collective themes, typically the following:

- Mine Automation and Robotics – automation suited to large bulk mining operation; robotics tailored more to smaller operations and customised activities. However, operational complexity, regulation and support requirements (labour and technology) would be drivers for partial automation, or operator assist. Whilst tele-remote operations are already well established, fully autonomous equipment is limited currently to mostly haulage and drilling, but has already been demonstrated to reduce variability, cut costs and increase efficiency.
- Data Integration, Predictive Analytics, Artificial Intelligence and Machine Learning. Coupled with the Internet of Things – assets and equipment will be remotely monitored in real-time to identify and, with Artificial Intelligence and machine learning, improve efficiency, safety, and reduce variability. Coupled with data diagnostics, asset or equipment condition can be used to anticipate failure or in the case of haul roads, distress, before it occurs, along with recommended preventive maintenance interventions or corrective actions.

These common themes all link into a hub of connected systems and processes which include remote and/or integrated operations centres, to provide the data and information necessary at multiple levels and across different functional and organisational boundaries of the business (Farrelly, 2016). This will also generate better end-to-end integration, and more focused and strategic partnerships to develop these new technologies and supply specific support activities. This would be based on the connectivity of people, processes and information from multiple data and sensors and specialised data analytics to provide insight and information virtually in any location.

Work by McKinsey (Durrant-Whyte et al., 2015), Deloitte (2017) and the World Economic Forum (2017) expand on how digital innovation especially could shape the future of mining. Allied to those changes would be the technologies and systems that allow for expansion and enrichment of the design, construction and especially performance monitoring aspects of mine road design. The CSIRO (2017) in a review of mining equipment, technology and services note that digital technology, data analytics and automation, systems connectively and mobility have the potential to improve productivity and driving disruptions across the mining value chain.

This chapter reviews some of the new technology and haulage equipment developments that stem from the advances towards mine automation, robotics, data integration, predictive analytics, artificial intelligence and machine learning as they may be applied to haul road design, construction, operation and management.

7.2 Autonomous haulage

In earlier chapters it was shown that a mine haul road network is a critical and vital component of the production process and this applies equally in autonomous (driverless) as well as conventional (human operator) truck-based hauling systems. In both cases, underperformance of the road will impact immediately on mine and fleet productivity and costs. Safety, productivity and equipment longevity are all dependent on well-designed, constructed and maintained haul roads.

As the concept of AHS moves from prototypes to production-ready applications, the operating performance of the haul road will become 'mission critical' to the overall success of autonomy in mining. Whilst in theory at least, an autonomous truck could operate on a much poorer road and would not be constrained by issues of operator comfort and the potential for strain injuries etc. associated with a poor road surface, the truck nevertheless rapidly accumulates strain damage and suffers from premature component failure as a result. Sensing road condition may be the key to minimising truck damage as speed could be adjusted to limit mechanical damage. Currently, AHS are based on conventionally design trucks and only as trucks evolve could consideration be given to operability and the option of accommodating harsher and significantly rougher road conditions.

As discussed by Thompson (2011b), with AHS, rapid deterioration in road performance will require costly remediation, human intervention and significant, albeit temporary, changes to operating procedures, to accommodate these types of events. As an example, with autonomous trucking, vehicle path wander is minimal and the road will be subject to high, channelised wheel loads over a limited area, without the wheel-path variations often encountered with conventional trucking, as can be seen in Figure 7.3. This is not dissimilar to what is currently experienced with trucks operating under a trolley assist system where wheel path wander is minimal due to the requirement to position the pantograph under the power lines.

This effect, coupled with the need for reliable and predictable performance requirements, presents challenges in mine road structural design, materials selection, performance specifications and construction. However, with autonomous trucks and channelised traffic comes the opportunity for instance, to reduce road construction and operating width, and therefore generate potentially significantly reductions in stripping ratios and improvements to mine economics, but only if the design of the road, and its associated deterioration rate, is predictable and manageable, based both on the materials used to construct the road and the maintenance (if any) required to be carried out on the road. Furthermore, high quality and strength pavement materials can be placed in the wheel tracks and poorer quality material outside the wheel tracks that are not trafficked, with concomitant construction and maintenance savings bearing in mind wind and water erosion outside the wheel tracks. Autonomous trucking has many potential advantages over conventional trucking, and to fully leverage these benefits, mine road design and management need to develop to address the requirements of autonomy in mining.

7.2.1 Autonomy technologies in mine haulage

Using AHS in an open pit improves safety, maintenance and equipment life, reduces fuel consumption and provides streamlined operations with increasingly accurate production systems. Parreira and Meech (2010) have estimated changes to a mine's Key Performance Indicators (KPIs) when this technology is adopted, as illustrated in Figure 7.4.

Looking further into the future of haulage units specifically, Albanese and McGagh (2011) anticipate an 'unconstrained' driverless vehicle, uncoupled from the requirements to house and inform a driver of the truck's location, operations and interactions in the mining environment. A truck that is more 'symmetrical', allowing multi-directional travel, all-wheel steering and drive systems with power- and energy-storage systems on-board under-body. This ideal combination of market-pull and manufacturer-push is leading to some of the new concept vehicles being either hypothesised or proposed currently, two examples of which are shown in Figure 7.5. A design of the right-hand image using batteries and self-generating power is on the drawing board, however, power and energy density of the battery systems is a

Figure 7.3 Typical channelisation associated with (top) autonomous trucks (Source www. komatsu.com/CompanyInfo/csr/environment/2013/pr-07.html) and trolley assist haulage (bottom).

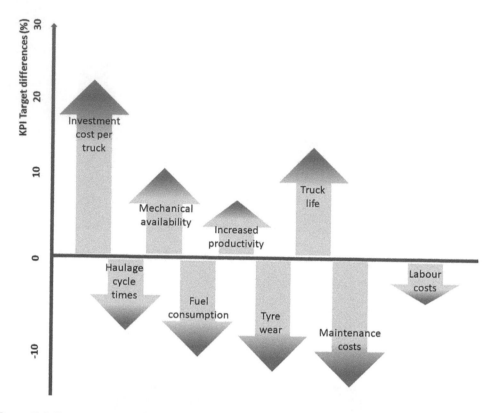

Figure 7.4 Estimated change to haulage KPIs associated with autonomous haulage trucks and associations to an assumed adequate haul road performance.

crucial design consideration in maintaining a realistic unladen vehicle mass (UVM):payload ratio. The Ciments Vigier SA modified Komatsu HD 605–7 (110t GVM) truck is an example, with the addition of electric drive and a 4.5 t, 700 kWh battery pack driving a synchronous electric motor capable of 590 kW of continuous power and 9500 Nm torque. The truck generated 40 kWh during a single (laden) descent and consumed a peak current of up to 3000 amperes while climbing (unladen) a 10% grade. The Artisan Z40 all-electric truck is a current development for underground haulage and showcases the potential of the technology to provide increases in peak power compared to a comparable diesel truck. In both cases, however, irrespective of the power system used, road condition and especially rolling resistance will remain critical to their cost-effective application, more so in the case where energy regeneration is used as the efficiency of the process will depend to a great extent on maximising energy input and minimising (rolling resistance) losses.

Much has been made of the concept and developmental requirements of the autonomous truck itself (Nebot, 2007), but to date little attention has been made to what that truck runs on, in terms of mine road design and management, and how this needs to evolve in tandem with the trucks themselves.

Figure 7.5 Conceptual haulage truck developments; Komatsu IAHV (top) and ETF Mining Truck (bottom).

Source: left, image courtesy of Komatsu Ltd; and right, image courtesy of ETF Holding d.o.o. Slovenia, ETF Equipment, 2018.

The decision to implement an AHS in a mine requires a thorough evaluation of impacts, not only operational improvements, but the inter-connectivity of this autonomy technology to other organisational processes. Lewis *et al.* (2004) noted that autonomous haulage has the potential to shatter the existing environment paradigms such as pit, haul road and equipment design. Parreira and Meech (2010) are analysing the behaviour of this new technology in terms of variations in adaptability and utilisation due to changes in various sub-systems or external environments, as a route to understanding the full interactions that may affect autonomous truck behaviour. Key amongst these agents/environments are haul road conditions and operating performance.

One of the major vehicle associated costs is that of tyres. Invariably tyres do not wear out, but are damaged by rocks cutting the tread or sidewall, or as a result of exceeding the limiting tyre TKPH (tonne-kilometre per hour). Inflatable tyres require pressures to be maintained to minimise damage to the truck and the tyre itself and under certain circumstances, failure of hot, pressurised tyres can be violent. Pressurised tyre technology is also a key consideration in the development of even larger trucks – where the tyre technology is unavailable, designers must resort to dual-wheel configurations with the attendant tyre changeout issues. Both manned and autonomous vehicles can be fitted with tyre pressure and temperature monitoring devices which, especially with autonomous systems and the associated control systems, allows potential tyre problems to be detected at an early stage. Non-pneumatic tyres are being patented to match the capabilities of the vehicle and Figure 7.6 shows an example of a US patent application, where the impacts applied to the tyre are absorbed by a rubber lattice sidewall (US Patent Application 13/954,504 dated 30 July 2013). The potential implication of a tyre of this nature on a haul road is fortunately minimal. If the tyre contact area is the same as the current pneumatic tyres, then for the same load the contact stress between tyre and road surface will be the same, and there will be no unexpected road damage. However, this should be evaluated for the particular tyre design adopted since as was explained in Chapter 3, in the majority of design problems, a circular tyre imprint is assumed and the contact radius determined as a function of tyre pressure (which in this case is not used).

7.2.2 Road design challenges for autonomous haulage and mining

The mine road design methodology shown in Chapter 1 Figure 1.5 forms the basis of the road design requirements and enhancements for mines in which AHS predominate. From the work of Hustrulid and Nillson (1998), Albanese and McGagh (2011) and Pukkila and Sarkka (2000), it is clear that an incremental approach is an appropriate means of meeting these challenges, and will allow mine road design technology to mature at the same rate as the AHS evolves. Thompson (2011b) described the haul roads technology enhancements required to meet these challenges. A near (1–3 years), intermediate (4–10 years) and long-term (11–20 years) research framework is required (in addition to on-going problem-solving 'tactical' research in mine road design and management):

- Near-term research is used to generate immediate solutions to design challenges that exist within the current available technology or systems, and the identification of longer-term, more complex issues which would inform intermediate- and long-term needs.

Figure 7.6 Non-pneumatic tyre for use on trucks and loaders.

- Intermediate-term research is needed to solve problems of greater complexity, identifying long-term opportunities fundamental systematic and technological changes.
- Long-term research is required to explore innovative solutions or far-reaching fundamental systematic and technological changes.

The emphasis of mine road design, construction and management research for the past 20 or more years has largely been on 'short-term' needs. These improvements, many of which have been described in the previous chapters, have not been matched by improvements in the fundamental technologies of mine road design and management, and although the quality of mine roads have probably increased as a result, many fundamental and longer-term issues have yet to be addressed: universal approaches to road life-cycle cost modelling, mine road-building material performance models, road management and performance optimisation, etc. Considerable research and innovation are required, from short-term tactical solutions to longer-term strategic systematic and fundamental technology and application improvements.

Increased mine truck traffic volumes and gross vehicle mass (GVM) has itself generated significant design challenges, especially when coupled with poor-quality road building materials (new surface mines in regions of thick regolith, weathered or transported material) and immature applications of potential remediating technologies (geo-fibres, geo-textiles, polymers, stabilisation, etc.). This has significantly increased pavement maintenance and rehabilitation costs, or, where these activities are absent, resulted in increased total costs. Construction practices and materials used in most mine roads will not provide adequate nor predictable performance in the presence of autonomous haulage systems. This situation will be aggravated if any of the environmental variables themselves (i.e. road geometry, vehicle speeds, GVM, materials, climate, etc.) depart further from what is known or experienced currently.

By analysing some of the key variables associated with autonomous haulage systems, a first indication of the road design future needs can be hypothesised. Current requirements such as low levels of dust because of safety considerations may no longer be a requirement unless it affects for example vehicle component durability or is an environmental constraint. Listed here are these generic needs and considerations, from near-term 'tactical' considerations, through to long-term needs of a more strategic nature.

- In the near-term, autonomous haulage systems deployment would require the following:
  - Assessment and documentation of 'as-built' conditions of the road, as a benchmark for performance and deterioration modelling
  - A real-time determination of road deterioration rates and locations
  - Condition-triggered maintenance interventions
  - Impact of channelised traffic on structural design requirements to be assessed
  - Rapid structural capacity determinations to be facilitated
  - Assessment of surfacing material functional performance limits
  - A mature understanding of the impact of climate on road performance and availability, and associated all-weather operating surface requirements

- In the intermediate-term, consideration should be given to these issues:
  - How to automatically gather and assess road performance and distress data survey
  - Using distress/condition survey data as basis for performance predictions and maintenance/rehabilitation planning

- Linking road performance data from various mine-sites to provide better regional predictive capabilities
- Developing a comprehensive approach to road rehabilitation design
- The impact of potentially steeper roads (>10% grade) on road and material performance
- The impact of narrower roads and channelised traffic on scheduled maintenance and rehabilitation
- Developing a better understanding of mine road life cycle costing and performance prediction techniques
- Relating road performance to truck damage, life-cycle costs and total road-user costs
- Evaluating the effect of maintenance strategies on road life, life-cycle costs and performance

- In the intermediate to long terms, the use of autonomous haulage systems in new un-explored geological domains, areas of thick regolith/weathered/transported material cover, with no road-building data or experience, is likely and would require the following:

  - Regional road-building material engineering characteristics data-base
  - Road performance and distress or degeneration rates models
  - Road life-cycle costing modelling (build, operate and maintain)

Finally, looking towards autonomous construction and maintenance/rehabilitation of the mine roads themselves, this would build on the near-, intermediate- and long-term goals and would additionally require the following:

- Strategies to asses in real-time and to modify or improve poor-quality road building materials (structural and functional design) specifications to align them with road-building requirements
- Rapid determination of 'as-placed' and 'as-built' conditions to determine correction construction (placement and compaction) requirements
- Interrogation of road performance and distress data to identify underlying design deficiency and appropriate remediation strategy
- Correlating road performance to design, construction and other factors, to improve predictive capability of performance and cost models

7.2.3 Roadmap for AHS haul road design and management

Figure 7.7 shows the broad generic challenges in road design development for AHS, from near-term 'tactical' considerations, through to long-term needs of a more strategic nature. The first theme is materials for road construction, design requirements, engineering specifications, performance limitations and strategies for improvements or enhancement of inherently poor or unsuitable materials. Allied to the materials themselves is the road management system aspects of maintenance and rehabilitation, specifically the requirement to predict and manage road maintenance and rehabilitation activities within the constraints of an AHS. Finally, total cost and life-cycle cost modelling combines with the other themes to enable an optimal design and operational strategy to be determined. Strategically, these themes would ultimately lead to a technology target specifying the road design requirements for autonomous

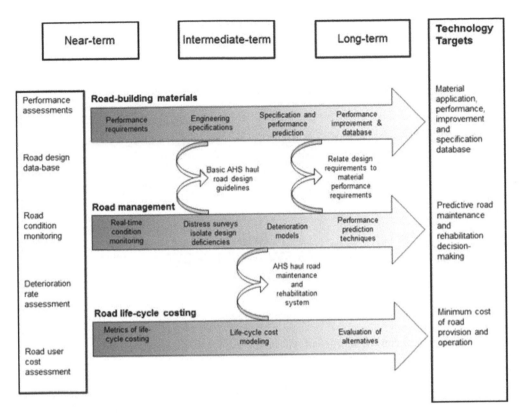

Figure 7.7 Roadmap for technology targets associated with autonomous haulage systems.

vehicle themselves, and, ultimately, to the autonomous mine in which road construction, management and maintenance or rehabilitation is achieved autonomously. Local conditions such as high phreatic surfaces may require changes to mining plans to permit operations under all conditions.

One of the oft-quoted benefits of AHS is associated with the highly channelised traffic which is accomplished using control systems that place the truck location accurately on the road. This leads to a potential reduction in road width, such that the road width geometrics covered in Chapter 2 can be re-specified to accommodate this greater positional accuracy. Initially, road width can be reduced to approach 2–2.5 times the width of a truck for two-way traffic, from the current 3.5W. The validity of this approach would require reference to the truck control systems, such that an oncoming vehicle is not seen as entering another trucks exclusion zone when it approaches to pass in the opposite direction. A benefit of reduced road width is that the pit size can be reduced or that less ore is sterilised as a result of the road width.

This modification can be further refined by designing the layerworks of the road specifically for laden and unladen routes, with a thinner structure being applied to the unladen segments of the haul network, as shown in Figure 7.8. Again, with the facility to specifically exclude all laden vehicles from the unladen carriageway, the adoption of this design approach is feasible.

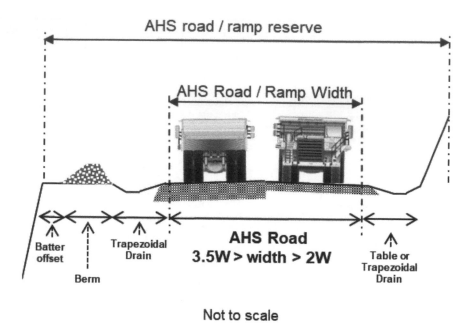

Figure 7.8 Potential reduction in road width with AHS applications.

A further development of this modification to established geometric design involves prediction of each truck's location within an AHS and the provision of a predominantly single-carriageway road, with pre-defined passing bays, such that priority rules favour laden haulage. In this case, the road width could be further reduced to one lane, as shown in Figure 7.9. As previously discussed, further benefit is accrued from reduced layerworks thickness associated with the passing bays for unladen vehicles only.

An optimisation exercise would be required to determine the appropriate number of passing bays, taking into consideration the length of haul route, traffic volumes, truck population, etc., such that overall delays in the system due to an enforced 'wait time' is reduced. In part, this could be readily achieved by dispatching unladen trucks at a specific speed or headway such that all vehicle interactions occur at one of the designated passing places. Again, with AHS, this level of control and predictability is possible, but as previously alluded to, would be dependent on the other aspects of road design, especially rolling resistance and road surface deterioration, to be addressed.

7.2.4 Premium or long-life pavements

As demonstrated in Chapter 6, there are significant financial benefits in providing a good structure and an appropriate wearing course, as this will reduce the maintenance costs such as grading and watering in addition to improving productivity of the haul trucks themselves. There is an international trend to provide a premium pavement through mixing in various chemical additives, with a view to realising the ultimate goal of any unpaved mine road – 24x7 trafficability. There is no general solution as the premium pavement has to be constructed

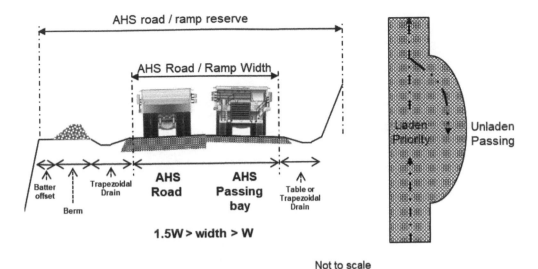

AHS road / ramp reserve

AHS Road / Ramp Width

Laden Priority   Unladen Passing

Batter offset

Trapezoidal Drain

Berm

AHS Road

AHS Passing bay

Table or Trapezoidal Drain

1.5W > width > W

Not to scale

Figure 7.9 Adoption of passing bays and single carriageway roads for AHS.

with locally available materials and the chemical additive has to be compatible and available. Experimentation as described in Chapter 4 would be required. The premium pavement is durable and produces a minimum of dust. All motor-grader maintenance is stopped as such activity would damage the hard crust, as well as water sprays, which cause erosion and potentially leaching of some additives. Since there is a minimum dust a rotary broom is effective to sweep the dust off the road without creating clouds of dust, which could be a safety hazard. An occasional diluted rejuvenation spray is applied to reinforce the surfacing. A careful loading procedure is applied to minimise spillage, and any spillage is removed either by rotary broom or by means of a hit-squad with a small rubber tyred loader.

Figure 7.10 shows a ramp road wearing course that has withstood three rainy seasons (1400 mm per year average rainfall) without any degradation or defects and a rolling resistance of 2% after three years. Traffic is about 100 kt per day. Minor abrasion has taken place.

The fact that the road surface is able to withstand the truck traffic has led to the hypothesis that it should be possible to construct a sealed road as on public roads. The maintenance regime for such a road is already in place. A sealed road should provide a life of at least 5 years, and probably longer, especially if modified binders are used. Special consideration must be given to the characteristics of the wearing course material to ensure that it fulfils the requirements of a base layer on a sealed road – a high clay content may attract moisture by means of capillary action, and this could lead to instability of the base, and cracking of the surfacing with resultant potholing. A dense graded crushed competent rock was used as wearing course in the example and this should meet the base material specifications.

7.3 Drone-based condition monitoring of haul roads

Drones (or UAVs) are used in a wide range of commercial roles ranging from search and rescue, surveillance, monitoring, firefighting, photography, videography and even delivery

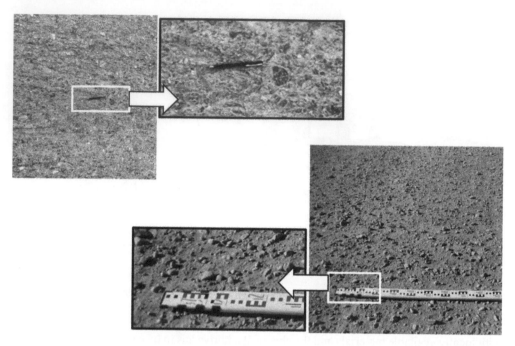

Figure 7.10 Close up of a premium pavement (LHS) after three rainy seasons, tightly bound showing little degradation, compared to an untreated pavement (RHS) with considerable unbound material as a result of loss of fines after rain.

services. Originally, drones were unmanned aircraft systems that were used to access locations that would be difficult or dangerous for a human. More recently, this has expanded to replacing or supplementing data gathering, irrespective of ease of access, as in many circumstances, their data gathering potential and efficiency is superior to 'boots on the ground'.

However, in support of this ability to quickly and efficiently capture data, the integration of drones and internet of things (IoT) technology is required to create enterprise use cases; drones working in conjunction with machine-linked and on-ground IoT sensor networks can help mining companies in numerous ways. By integrating UAVs with on-ground equipment, Komatsu (2016) use UAV data and artificial intelligence to manage construction sites and equipment almost fully autonomously. The system combines Komatsu's intelligent Machine Control (iMC) technology with drone-generated data to automate (initially foundation) and other civil construction work, but could easily be expanded to mine haul road construction and, ultimately, maintenance. Time-of-Flight depth ranging camera sensors are used for object scanning, navigation, obstacle avoidance, feature recognition, volumetric measurement and 3D photography. Other UAV sensor platforms cover multispectral, lidar, photogrammetry and thermal vision sensors that can be used in mining to construct 3D models of the mining environment, including roads, and, with increasing resolution, the condition of the road surface itself can also be evaluated autonomously. Examples of these applications are described by Airware (2018) (Figs. 7.11–7.13), Propeller Aero (2018) (Figs. 7.14 and 7.15) and Airobotics (2018) amongst others.

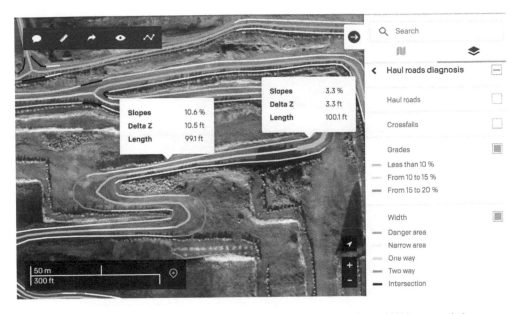

Figure 7.11 Example of haul road width and grade monitoring from UAV-sourced data.
Source: image courtesy of Airware.

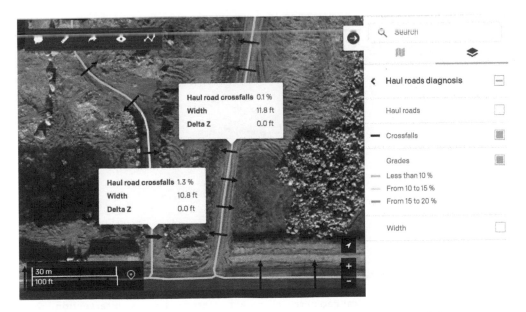

Figure 7.12 Example of haul road crossfall measurements from UAV-sourced data.
Source: image courtesy of Airware.

Figure 7.13 Example of haul road hazard identification from UAV-sourced data.
Source: image courtesy of Airware.

Figure 7.14 Example of vertical geometric alignment assessment across intersection.
Source: image courtesy of Propeller Aero.

Figure 7.15 Example of safety berm measurement (angle of inner face and height).

Source: image courtesy of Propeller Aero.

Airware provides a haul road module to automatically extract road condition parameters to provide a basis for haul road maintenance or improvement. In its simplest form, road boundaries, widths, grades, cambers, etc. are determined. When overlain on a digital aerial survey of the site, it enables problem areas to be identified and serves as a common data platform for further analysis and more-informed decision making. Specifically, the ability to identify uneven, degraded surfaces that contribute to higher vehicle operating costs and safety hazards, as well as areas of poor drainage that would result in ponding of water, tyre damage and more rapid deterioration. Volumetric surveys can also be applied to monitor the road construction process, including layerworks thickness. As sensor platform technology develops, together with the research fundamentals that support the analytics, it would eventually become possible to both measure and manage the construction process fully, including material selection, laydown, compaction etc.

Data from UAVs can also be leveraged from the current condition assessment, towards defect recognition and remote assessment of rolling resistance (using the defect rating system described in Chapter 5). This would be artificial intelligence–based and as such would require training in defect recognition in order for the system to reliably interpret data and classify individual defects degree ratings. Defect extent ratings, however, would be well within the capacity of current system analytics.

7.4 Haul road geometrical diagnostics – an example

The aerial survey produces, with photogrammetry, an ability to generate a dense point cloud after processing the data captured by the on-board camera in the aircraft. Figure 7.16 presents the results of the survey in a projected view of the point cloud coloured from the orthophoto created.

Figure 7.16 3D view of the road surveyed with photogrammetry.

From the dense cloud a digital elevation model (DEM) can be created as well as contour plots at 20 cm spacing. From these models it is possible to generate cross sections (longitudinal and transverse) to analyse the road profile and the general geometrical state of the road.

Typical cross sections numbered as shown in Figure 7.17 can be used to perform the geometrical analysis from the data captured by the aerial survey:

- Section 1 – Longitudinal straight section approximately along the road axis
- Section 2 – Longitudinal curved section approximately along the road axis
- Section 3 – Longitudinal straight section approximately along the secondary road axis
- Section 4 – First cross section in the main road
- Section 5 – Second cross section in the main road

The detail of the access along Section 1 is shown in Figure 7.18. It can be seen that there is an average road gradient of approximately 6%, as shown by the red dotted line. The red dotted line starts from a horizontal reference of 20 m and an elevation of 1275 m and extends to the point 220 m with an elevation of approximately 1289 m. Thus the vertical distance difference is 14 m for a horizontal distance difference of 220 m, resulting in 14/220 = 0.058 or 5.8% overall slope. However, it can be seen that the stretch between these two points is slightly undulating longitudinally. This undulation can mean a source of variation in the speed of the trucks and should be avoided by systematic monitoring during the construction of the roads. This kind of mapping survey can also be done by conventional land survey teams, but it is important to note that with the aerial vehicle, the coverage area is larger, the results are faster and also safer for the surveyors that do not need to walk along the roads

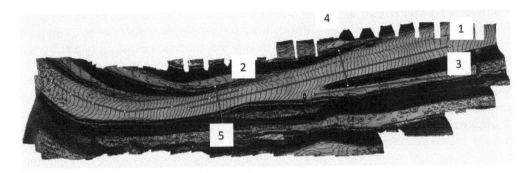

Figure 7.17 Orthomosaic with contours raised from the stretch of road.

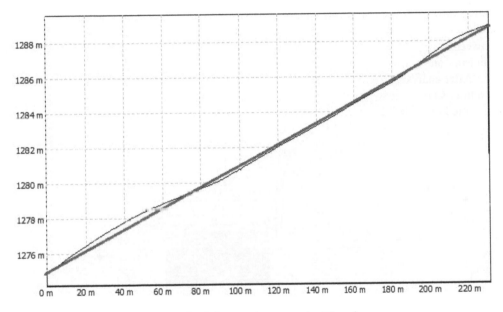

Figure 7.18 Longitudinal section I of the road surveyed with a drone.

while the trucks and all the mobile equipment are operating. Note that a survey of a 80 hectare pit took an hour to fly and an afternoon to process. This would be a task that would take a survey team at least 3 months.

There are two ways in which the quality of the road surface can be evaluated by means of a drone survey, namely by visual or machine learning evaluation of the photographic evidence, or by means of computing the riding quality. This information may be used to do the following:

i.   assess the performance of a road, and
ii.   assess the impact of the surface road on the operating costs of vehicles using the road.

Both techniques are still in their infancy, but they show potential. Visual assessment of road quality is a slow process and not yet suited for real-time implementation. The International Roughness Index (IRI) is used on public roads and laser elevations are processed by a computer package (Sayers and Karamihas, 1998) that simulates the vertical accelerations of a passenger vehicle with standardised characteristics, the so-called golden car. It was possible to generate the IRI from drone data, but currently the resolution and repeatability of drone measurements in the vertical axis is about 30 mm, whereas the profile resolution has to be less than 2 mm. Once more precise and higher resolution data becomes available this approach can be applied.

Figure 7.19 shows the cross section related to section line 5 in Figure 7.17. In the horizontal axis the distances are measured in meters and in the vertical axis the actual elevations are also presented in meters. On the right-hand side, a 2 m-high safety berm is observed, which corresponds to a suitable size for the largest truck in operation (Komatsu 730E). The equipment, according to the manufacturer's recommendation, uses 37.00R57 tyres, and at the appropriate inflation pressure, has a diameter of 3.42m. The actual berm height in this case corresponds to the recommended berm height for an effective protection for runaway trucks (although the inner face of the berm should ideally be steeper as discussed in Section 2.6). The transverse slope directing the drainage to the inner side of the road way has a slope of approximately 1.6%. The recommended slope is between 2% and 3% to promote faster drainage of surface water to the roadside drains with greater efficiency and speed (but still minimising scour), especially during periods of higher rainfall.

After analysing the main geometric elements detected in the survey, the actual measured conditions are compared with the mine's recommended standards for this type of project, as presented in Table 7.1.

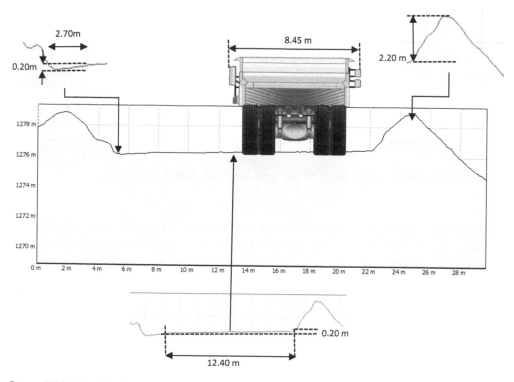

Figure 7.19 Details of the geometrical elements at transverse section 5.

Table 7.1 Description of the geometric elements taken from the survey.

Geometry Checklist	Measured	Recommended	Checklist finding
Berm height (m)	2.20	2.26	CORRECT
Road (trafficked) width (m)	12.40	3.5 × W = 29.6	Inadequate
Longitudinal grade	6%	MÁX 10%	CORRECT
Transverse slope	1.6%	2–3%	Inadequate
Side drain (H × W in m)	0,2 × 2,7	0,4 × 2,0	
Cross sectional area of drain (m²)	0,27	0,40	Inadequate

It should be noted that the cross-sectional area of the drain is somewhat dependent on the runoff area of a given portion of the road, which means a dependence on the width of the road, the internal and peripheral drainage system. For example, if there are water withdrawals through side outlets within regular intervals that area may vary according to those intervals. But what is advocated here is that this drainage area would be satisfactory for a design width appropriate to the equipment (two lanes for KOMATSU 730E) with a road width and accesses that have systematic withdrawals every 50–70 m.

In cases where road conditions, alignments, drainage, traffic signage and similar on-road features are sought, smartphone hosted open-source applications such as Mapillary can be usefully applied to map and record road conditions for further processing, image or feature recognition or simply for road maintenance diagnostics. Figure 7.20 shows a typical application in which a section of a road network has been recorded, geo-located on a map of the location or mine site and then subsequently used in this case to assess variations in elevation and thus grade of the road. The approach can have further application in road condition monitoring by applying the functional assessment methodology discussed previously in conjunction with machine-learning image recognition techniques to both identify critical functional defects and, ultimately, determine rolling resistance.

7.5 Fleet management systems

Fleet management systems, commonly known as dispatch systems, evolved from the 1970's from simple applications with little use of computational resources. From the very beginning, using manual batch allocation, passing through radio communication to direct and control equipment from source to destination, to the latest modern online telemetry, real time positioning, speeds and traffic providing various information which is received in the mine control office or data centre. The data centre might be located remotely and operated by a reduced number of people for the entire fleet, making the operation react immediately by communicating directly with the operator or mine supervision.

A fleet (vehicle) management system can include a range of functions, alerts and triggers according the specific need of each operation. Fleet management data can include vehicle operating hours, vehicle telemetry (tracking and diagnostics), driver performance, speed control, fuel consumption and also health and safety management. The system facilitates identifying improvements in efficiency, productivity and the opportunities for risk reduction associated with vehicle investment, for haulage fleet operators.

Several data collected by a dispatch system can be used to improve the performance not only of the truck fleet and mining equipment in general, but it can also generate information

Figure 7.20 Example of road diagnostic recording and analysis using Mapillary software.

to detect haul roads problems and indicate the need and prioritisation of corrective actions for the maintenance crew, as part of an integrated haul road management system.

Information on truck speed, vertical acceleration, un-planned stops during the operating cycle, etc. can indicate problems with the road or can even be used to help to back calculate rolling resistance for certain areas. As can be seen in Figure 7.21, the number of events from a single shift of operation can be enormous. Consider for example a fleet with 20 trucks recording events at every 10 s, this represents around 57,600 events registered in a single shift of 8 h across all mine roads, origins and destinations, etc., Figure 7.22 shows travel speeds events received from a small section of a haul road network over one shift of operation. Sometimes the number of events can be so large that the practicality of treating this data to generate information and then understand the problems in real time or even to react in-shift to correct the defects related to the roads can be difficult without a good control system. Furthermore,

Figure 7.21 Events from a dispatch system for a shift of work.

Figure 7.22 Events collected from the truck fleet by the dispatch management system.

highly trained and experienced staff are required to both recognise the appropriate and relevant data and to generate action plans. It is really necessary to integrate and filter the relevant information to analyse and provide feedback to the operation as sometimes the systems are not ready to deal with the specific routines to compile, treat and analyse data.

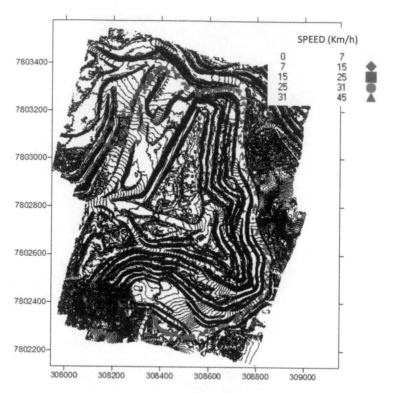

Figure 7.23 Mine layout showing variation of laden truck speeds along specific haul roads.

The mine plan and the current mine layout can be updated daily to reflect the real status of the mine while the equipment is operating, and this updated data used on a daily basis to support the concept of a 'real time' mine. Figure 7.23 shows the speed records for a single operator on laden hauls over the whole shift. It can be seen that there are repeated patterns of speeds along the various stretches of the road. This information and especially the variation and rate of change in speeds could form the basis of a road condition assessment and a maintenance intervention value-proposition. Nowadays in high-production, low-cost operations it is almost unthinkable to attempt to manage and optimise operational performance without a dispatch system in place, forming the backbone of the operation's data collation and analytics.

References

Addo, J.Q. & Sanders, T.G. (1995) Effectiveness and environmental impact of road dust suppressants. MPC Report 94-28A, Mountain Plains Consortium, Fargo, ND, USA.

Ahlvin, R.G., Ulery, H.H., Hutchinson, R.L. & Rice, J.L. (1971) Multiple wheel heavy gear load pavement tests. Vol. 1 Basic Report, USA Waterways Exp. Stn Report, AFWL-TR-70-113.

Airobotics (2018) *Benefits of Automated Drones for Haul Road Optimization.* Available from: www.airoboticsdrones.com/applications/haul-road-optimization/ [accessed April 2018].

Airware (2018) *Haulroad Module.* Available from: www.airware.com/en/industries/mining-quarrying/haul-road-module/ [accessed April 2018].

Albanese, T. & McGagh, J. (2011) Future trends in mining. In: SME Mining Engineering Handbook (ed) Darling. Society for Mining, Metallurgy and Exploration, INC (SME), Littleton, CO, USA.

Aldinger, J.A., Kenney, J.M. & Keran, C.M. (1995) *Mobile Equipment Accidents in Surface Coal Mines.* U.S. Department of the Interior Bureau of Mines IC 9428, Washington, DC, USA.

Amponsah-Dacosta, F. (1997) Cost-effective strategies for dust control in an opencast coal mine. MSc Project Report, University of the Witwatersrand, Johannesburg, South Africa. pp. 27–48, 53–57.

Angell, D.J. (1988) *Technical Basis for the Pavement Design Manual.* Main Roads Department, Brisbane, QLD, Australia.

AusIMM (2012) *Cost Estimation Handbook*, 2nd ed. – Monograph 27. The Australasian Institute of Mining and Metallurgy, Carlton, Vic., Australia.

Australian Roads Research Board-ARRB (1996) In road dust control techniques, evaluation of chemical dust suppressants' performance. Australian Roads Research Board Special Report 54, Victoria, Australia. pp. 17–18.

Australian Transport Council (ATC) (2006) *National Road Safety Action Plan 2007–2008.* Australian Transport Council, Canberra, ACT, Australia.

AustRoads (1992) *Pavement Design: Guide to the Structural Design of Road Pavements.* AP 17/92, AustRoads, Sydney, NSW, Australia.

AustRoads (2009a) *A Review of Relationship to Predict Subgrade Modulus From CBR: National Association of Road Transport and Transport Authorities of Australia.* AustRoads Publication AP-T130/09, Sydney, NSW, Australia.

AustRoads (2009b) *Guide to Road Safety -Part 6: Road Safety Audit.* AustRoads Incorporated, AGRS06/09, Sydney, NSW, Australia. ISBN 1 921551 10 9.

Bowles, J.E. (1984) *Physical and Geotechnical Properties of Soils*, 2nd ed. McGraw-Hill Book Company, New York, NY, USA.

Boyd, W.K. & Foster, C.R. (1950) Design curves for very heavy multiple wheel assemblies. In development of CBR flexible pavement design methods for airfields. A Symposium. *Transactions of the American Society for Civil Engineering*, 115, paper 2406, 534–546.

Brown, M., Mercier, S. & Provencher, Y. (2002) Road maintenance with Opti-Grade®. *Eighth International Conference on Low Volume Roads, Journal of the Transportation Research Board*, TRR 1819,

Vol. 1, Transportation Research Board Research Board of the National Academies, Washington, DC, USA. pp. 282–286.

Caterpillar Inc. (1999) *VIMS Vehicle Information Management System*. CD-ROM, EFRK 0060, Caterpillar Inc., Peoria, IL, USA.

Caterpillar Inc. (2000) *Caterpillar Road Analysis Control Salesgram. Construction and Mining Trucks MCE*. Caterpillar Inc., Peoria, IL, USA.

Caterpillar Inc. (2015) *Caterpillar Performance Handbook Edition 45*. Caterpillar Inc., Peoria, IL, USA.

Caterpillar Inc. (2017) *How Haul Road Maintenance Can Extend Tire Life*. Available from: www.cat.com/en_US/by-industry/mining/articles/haul-road-maintenance.html

Chesher, A. & Harrison, R. (1987) *'Vehicle Operating Costs – Evidence from Developing Countries', Highway Design and Maintenance Standards Series*. Transportation Department, World Bank, Washington, DC, USA.

Committee of State Road Authorities (CSRA) (1987) Technical Recommendations for Highways. TRH8 Design and use of hot-mix asphalt in pavements, Draft TRH8, Appendix A, Pretoria, South Africa.

Committee of State Road Authorities (1990) Draft TRH 20 The structural design, construction and maintenance of unpaved roads. Department of Transport, Pretoria, South Africa. ISBN 0 908381 875.

Cooper, A. (2008*) Modelling Truck Performance in a Spreadsheet. Principal Mining Consultant, Snowden Mining Consultants. Internal Document*, Perth, WA, Australia.

Dunston, P.S., Sinfield, J.V. & Lee, T. (2007) Technology development decision economics for real-time rolling resistance monitoring of haul roads. *Journal of Construction Engineering and Management*, 133(5), 393–402.

Durrant-Whyte, H., Geraghty, R., Pujol, F. & Sellschop, R. (2015) *How Digital Innovation Can Improve Mining Productivity*. Available from: www.mckinsey.com/industries/metals-and-mining/our-insights/how-digital-innovation-can-improve-mining-productivity [accessed April 2018].

ETF Equipment (2018) *ETF Mining Trucks*. Available from: http://etf.equipment/ [accessed 7 October 2018].

Erarslan, K. (2005) Modelling Performance and Retarder Chart of Off-Highway Trucks by Cubic Splines for Cycle Time Estimation, Mining Technology (Trans. IMM-Section A), 114, September, A161-A166.

Farrelly, C.T. (2016) *Are We Ready for the Brave New World in Mining?* Available from: http://indago partners.blogspot.com.au/2016/03/are-we-ready-for-brave-new-world.html [accessed April 2018].

FHWA (1985) *ELSYM5A: Interactive Version 5A Users Guide*. FHWA, US Department of Transportation, Washington, DC, USA.

Fourie, G.A.F., Smith, R.A.F., Thompson, R.J. & Visser, A.T. (1998) Benchmarking haul road design standards to reduce transportation accidents. *International Journal of Surface Mining and Reclamation Engineering*, 12(4), December, 157–162.

Fricke, L.B. (1990) *Traffic Accident Reconstruction: The Traffic Accident Investigation Manual*, Vol. 2. Northwestern University Press, Evanston, IL, USA.

Gilewicz, P. (2006) The market for large mining trucks & loading equipment: What's happened & what's ahead. *CD-ROM. Mining Media Haulage and Loading Conference*, 24–26 May, Phoenix, AZ, USA.

Gleisner, S., Albornoz, J., Graell, T. & Halles, F. (2017) The economic impact of haul road maintenance strategies in the success of long-term mine plans. *Proceedings 5th International Seminar on Mine Planning, Santiago, Chile, 23–25 August*.

Goodyear (2009) *Off the Road – Tire Maintenance Manual – Digital Copy*. Available from: www.goodyearotr.com/resources/engineering-data [accessed 17 May 2018].

Goodyear (2017) *Off the Road Tires – Engineering Data*, updated in 31 August 2017 digital copy. Available from: www.goodyearotr.com/resources/engineering-data [accessed 17 May 2018].

Hawkey, N.T. (1982) Haul road management techniques. *Australian IMM Conference on Off-Highway Truck Haulage*, Mt Newman, Australia. pp. 15–23.

Huang, Y.H. (1993). *Pavement Analysis and Design*. Prentice Hall, NJ., USA.

Hugo, D., Heyns, P.S., Thompson, R.J. & Visser, A.T. (2008) Haul road condition monitoring using vehicle response measurements. *Journal of Terramechanics*. doi:10.1016/j.jterra.

Hunting, K.L. & Weeks, J.L. 1993. Transport injuries in small coal mines: An exploratory analysis. *American Journal of Industrial Medicine*, 23, 391–406.

Hustrulid, W.A. & Kuchta, M. (2006) *Open Pit Mine Planning and Design*, 2nd ed. Taylor and Francis, London, UK.

Hustrulid, W.A. & Nilsson, J.O. (1998) Automation and productivity increases at LKAB, Kiruna, Sweden. *Proceedings of CMMI Congress 1998*, Montreal, Canada.

InfoMine (2017) *Mine & Mill Equipment Costs: An Estimator's Guide, Electronic Edition*. InfoMine Inc, Vancouver, Canada.

Ingle, J.H. (1991) Good tyre management program cuts costs and downtime. *Coal Magazine*, 8(6), 113–116.

Jones, D. (1996) The impacts and control of dust on mine haul roads. In: Glen, H.W. ed., *Proceedings of the South African Institute of Mining and Metallurgy Conference on Surface Mining*, 30 September–4 October 1996, Sandton, South Africa. pp. 351–355.

Jones, D. (1999) *Holistic Approach to Research into Dust and Dust Control on Unsealed Roads*. Division of Roads and Transport Technology, Council for Scientific and Industrial Research, Pretoria, South Africa.

Jordaan, G.J. (1994) The South African mechanistic pavement rehabilitation design method. Research Report RR91/242, Department of Transport, Pretoria, South Africa.

Joseph, T.G., Curly, M. & Anand, A. (2017) Operational methodologies for rolling resistance evaluation. *Geotechnical and Geological Engineering*, 35(6).

Kaufman, W.W. & Ault, J.C. (1977) *Design of Surface Mining Haulage Roads – a Manual*. U.S. Department of Interior, Bureau of Mines, Information Circular 8758.

Kecojevic, V.J., Komljenovic, D., Groves, W. & Radomsky, M. (2007) An analysis of equipment related fatal accidents in US mining operations: 1995–2005. *Safety Science*, 45, 864–874.

Kecojevic, V.J. & Radomsky, M. (2004) The causes and control of loader and truck-related fatalities in surface mining operations. *Injury Control and Safety Promotion*, 11(4), 239–251.

Komatsu (2016) *Intelligent Machine Control*. Form No: ZESB099913_November2016. Available from: www.komatsu.com.au/AboutKomatsu/iMC_Brochure_final_webready_V2.pdf [accessed 1 August 2018].

Lea, J.D. & Jones, D.J. (2007) Initial findings on skid resistance of unpaved roads. *Transportation Research Record: Journal of the Transportation Research Board*, 2016, Transportation Research Board of the National Academies, Washington, DC, USA. pp. 49–55.

Lee, T.Y. (2010). Development and Validation of Rolling Resistance-Based Haul Road Management. Purdue University Graduate School Thesis, West Lafayette, IN, USA.

Lewis, M.W., Werner, J. & Sambirsky, B. (2004) Capturing unrealised capacity. *CIM Bulletin*, 97(1076), ProQuest Science Journals, Montreal, Canada, 57–62.

Long, G., 1968. 'Road and property maintenance', Surface Mining, (ed. Pfleider, E.P), Seeley W Mudd Series, American Institution of Mining, Metallurgical and Petroleum Engineers, Inc., New York, NY, USA. pp. 678–682.

MainRoads, W.A. (2013) *Engineering Road Note 9, Materials Engineering Branch*. MainRoads, Perth, WA, Australia. Available from: www.mainroads.wa.gov.au/BuildingRoads/StandardsTechnical/MaterialsEngineering/Publications/Pages/Engineering_Road_Notes.aspx [accessed 20 May 2018].

Miller, R.E., Thompson, R.J. & Lowe, N.T. (2004) A GPS-based system for minimizing jolts to heavy equipment operators. *Society of Automotive Engineering, SAE Transactions, Journal of Commercial Vehicles*, SAE 2004-01-2726, ISBN 0-7680-1641-X. pp. 850–855.

Mine Health and Safety Council (MHSC) (2001) Influence of road design and construction on transport accidents. Report ORT 308. Safety in Mines Research Advisory Committee. Final Project Reports CD, Vol. 1, 1994–1996, Johannesburg, South Africa.

Minerals Council of Australia (MCA) (2006) *The Mobile Equipment Incident Causation Survey (Meics) 2005–2006 Survey Findings and Recommendation*, MCA, Melbourne, Vic., Australia.

Morgan, J.R., Tucker, J.S. & McInnes, D.B. (1994) Mechanistic design approach for unsealed mine haul roads. *Proceedings of the 17th ARRB. Conference. Part 2 (of 7)*. ARRB. pp. 69–81.

Nebot, E.M. (2007) Surface mining: Main research issues for autonomous operations. *Robotics Research*, STAR publication No. 28, Springer-Verlag, Berlin, Germany. pp. 268–280.

Nogami, J.S. & Villibor, D.F. (1991) Use of lateritic fine-grained soils in road pavement base courses. *Geotechnical and Geological Engineering*, 9, 167. https://doi.org/10.1007/BF00881739

Otte, E. (1979) The CBR pavement design method (Die KDV – plaveiselontwerpmetode). *Die Siviele Ingenieur in Suid-Afrika*, 21(4), 87–93.

Paige-Green, P. (1990) Some surface roughness, loss and slipperiness characteristics of unpaved roads. In: Meyer, W.E. & Reichert, J. (eds) *Surface Characteristics of Roadways: International Research and Technologies, ASTM STP 1031*. American Society for Testing and Materials, Philadelphia. pp. 268–291.

Parreira, J. & Meech, J. (2010) Autonomous vs manual haulage trucks – how mine simulation contributes to future haulage system developments. *CIM Meeting 2010*, Vancouver, BC, Canada.

Paterson, W.D.O. (1987) *Prediction of Road Deterioration and Maintenance Effects: Theory and Quantification. The Highway Design and Maintenance Standards Study*, Vol. 3. Transportation Department, World Bank, Washington, DC, USA.

Perdomo, J.L., (2001) Detailed Haul Unit Performance Model. Virginia Polytechnic Institute and State University, Blacksburg, VA, USA.

Powell, W.D., Potter, J.F., Mayhew, H.C. & Nunn, M.E. (1984) The structural design of bituminous roads. Laboratory Report 1132, Transport and Road Research Laboratory, Crowthorne, UK.

Proof Engineers (2017) *Road Condition Monitoring*. Available from: www.proofengineers.com.au/systems/road-condition-monitoring [accessed 7 October 2018].

Propeller Aero (2017) *Six Smart Ways Mining and Aggregates Businesses are Using Drones*. Available from: www.propelleraero.com/blog/six-ways-mining-and-aggregates-businesses-use-drones/ [accessed April 2018].

Pukkila, J. & Sarkka, P. (2000) Intelligent mine technology program and its implementation. *Proceedings of Massmin 2000*, Brisbane, OLD, Australia.

Queensland Government Department of Transport and Main Roads. ARNDT v1.03.06. Available from: www.tmr.qld.gov.au/business-industry/Road-systems-and-engineering/Software/ARNDT [accessed 7 October 2018].

Randolph, R.F. & Boldt, C.M.K. (1996) Safety analysis of surface haulage accidents. *Proceedings of 27th Annual Institute on Mining Health, Safety and Research*. Virginia Polytechnic Institute and State University, Blacksburg, VA, USA. pp. 29–38.

Rio Tinto (2018) *Mine of the Future*. Available from: www.riotinto.com/australia/pilbara/mine-of-the-future-9603.aspx [accessed April 2018].

Sampson, L.R. & Netterberg, F. (1985) The cone penetration index: A simple new soil index to replace the plasticity index. *Proceedings of 11th International Conference on Soil Mechanics and Foundation Engineering, 12–16 August*. San Francisco. pp. 1041–1048.

Sayers, M.W. & Karamihas, S.M. (1998) *The Little Book of Profiling – Basic information about Measuring and Interpreting Road Profiles*. University of Michigan., The Regent of University of Michigan, Ann Arbor, MI, USA.

Sharma, B. & Bora, P.K. (2003) Plastic limit, liquid limit and undrained shear strength of soil – reappraisal. Available from: ascelibrary.org/doi/full/10.1061/(ASCE)1090-0241(2003)129:8(774) [accessed 27 October 2017].

Simpson, G.C, Rushworth, A.M, Von Glehn, F.H. & Lomas, R.H. (1996) Investigation into the causes of transport and tramming accidents on mines other than coal, gold and platinum. Mine Health and

Safety Council Safety in Mines Research Advisory Committee, Other Mines Sub-committee final project report OTH202, Department of Minerals and Energy, Pretoria, South Africa.

SRS1 Software (2017) Available from: www.srs1software.com/SRS1CubicSplineForExcel.aspx [accessed 10 November 2017].

Tannant, D.D. & Regensburg, B. (2001) *Guidelines for Mine Haul Road Design.* Edmonton, Canada: School of Mining and Petroleum Engineering, University of Alberta. Available from: www.smart mines.com/haul roads/guidelines.pdf [accessed 12 May 2018].

Thompson, R.J. (1996) *The design and maintenance of surface mine haul roads.* PhD Thesis, University of Pretoria, South Africa.

Thompson, R.J. (2011a) Design, construction and management of haul roads. In: SME Mining Engineering Handbook (ed) *Darling.* Society for Mining, Metallurgy and Exploration, INC (SME), Littleton, CO, USA. pp. 957–976.

Thompson, R.J. (2011b) Mine road design and management in autonomous hauling operations: A research roadmap. *Paper 21, AUSIMM Second International Future Mining Conference,* Sydney, NSW, Australia, 22–23 November.

Thompson, R.J. (2017) Can big data answer the big question – how do my haul roads perform? *Mining Media International Haulage and Loading Conference,* 7–10 May, Phoenix, AZ, USA.

Thompson, R.J. & Visser, A.T. (1996) Towards a mechanistic structural design method for surface mine haul roads. *Journal of the South African Institute of Civil Engineering,* 38(2), Second Quarter, Johannesburg, South Africa.

Thompson, R.J. & Visser, A.T. (1999) *Management of Unpaved Road Networks on Opencast Mines: Transportation Research Record (TRR) 1652,* Transportation Research Board Research Board of the National Academies, Washington, DC, USA. pp. 88–97.

Thompson, R.J. & Visser, A.T. (2000a) The functional design of surface mine haul roads. *Journal. of the South African Institute of Mining and Metallurgy,* 100(3), May/June, 169–180.

Thompson, R.J. & Visser, A.T. (2000b) The reduction of the safety and health risk associated with the generation of dust on strip coal mine haul roads. Mine Health and Safety Council Safety in Mines Research Advisory Committee, Collieries Sub-committee Final Report for Project COL 467, Pretoria, South Africa. pp. 34–41, 117–141.

Thompson, R.J. & Visser, A.T. (2002) *Benchmarking and Managing Surface Mine Haul Road Dust Emissions.* Transactions of the Institute of Mining and Metallurgy (UK), Section A. p. 113.

Thompson, R.J. & Visser, A.T. (2003a) Mine haul road fugitive dust emission and exposure characterisation. *WIT Conference: 2nd International Conference on The Impact of Environmental Factors on Health, 11–13 September.* Catania, Sicily, Italy. pp. 103–112.

Thompson, R.J. & Visser, A.T. (2003b) Mine haul road maintenance management systems. *Journal. of the South African Institute of Mining and Metallurgy,* 103(5), May/June.

Thompson, R.J. & Visser, A.T. (2006a) The impact of rolling resistance on fuel, speed and costs. HME 2006: Continuous improvement case studies.

Thompson, R.J. & Visser, A.T. (2006b) Selection and maintenance of mine haul road wearing course materials. *Transactions of the Institution of Mining and Metallurgy, Section A: Mining Technology,* 115(4), 140–153.

Thompson, R.J., Visser, A.T., Heyns, P.S. & Hugo, D. (2006) Mine road maintenance management using haul truck response measurements. *Transactions of the Institution of Mining, Metallurgy, Section A: Mining Technology,* 115(4), 123–128.

Thompson, R., Visser, A.T., Miller, R. & Lowe, T. (2003) Development of real-time mine road maintenance management system using haul truck and road vibration signature analysis. *Transportation Research Record,* 1819(1), 305–312.

Thompson, R.J, Visser, A.T., Smith, R.A.F. & Fourie, G.A.F. (1997) Final report for project SIMOT 308. Mine Health and Safety Council Safety in Mines Research Advisory Committee, Other Mines Sub-committee, Department of Minerals and Energy, South Africa.

Tulloch, D. & Stocker, D. (2011) *Coal Mine Road Network Surface Friction Report*. Road Safety Training Services Pty Ltd. Available from: http://www.qldminingsafety.org.au/?page_id=807

Turnbull, W.J. & Ahlvin, R.G. (1957) Mathematical expression of the CBR relationships. *Proceedings of 4th International Conference on Soil Mechanics and Foundation Engineering*, Vol. 2, London, UK. p. 178.

United States Bureau of Mines (USBM) (1981) *Study of Mine Haulage Roadway Surface Safety Hazards*. U.S. Department of the Interior, Bureau of Mines. Mines, Minerals, Health and Safety Technology Division, Bumines, OFR. pp. 5–83.

United States Environmental Protection Agency -USEPA (1995) *In National Air Pollution Emission Trends, 1990–1995*. Office of Air Quality Planning and Standards, United States EPA/454/R-96-007, Research Triangle Park, NC, USA. pp. 231–239.

USDOI United States Bureau of Mines (USBM) (1981) *Study of Mine Haulage Roadway Surface Safety Hazards*. U.S. Department of the Interior, Mines, Minerals, Health and Safety Technology Division, Bumines, OFR. pp. 5–83.

Vagaja, D. (2010) Road safety audits on mining operations. *Queensland Mining Industry Health and Safety Conference*, Townsville, QLD, Australia.

Visser, A.T. (1981) *An evaluation of unpaved road performance and maintenance*. PhD dissertation. The University of Texas at Austin, May 1981.

Visser, A.T. & Erasmus, F. (2005) Practical evaluation of additives used for soil stabilisation. Institute of Municipal Engineers in South Africa, 30(4), April.

Weinert, H.H. (1980) *The Natural Road Construction Materials of Southern Africa*. Academia, Pretoria, South Africa.

Western Mine Inc. (2002) *Mining Cost Service*. Rodrigo, Spokane, Washington, DC, USA.

World Economic Forum (2017) *Digital Transformation Initiative: Mining and Metals Industry (In Collaboration with Accenture)*. REF 060117, World Economic Forum, Geneva, Switzerland.

Yoder, E.J. & Witczak, M.W. (1975) *Principles of Pavement Design*. John Wiley & Sons, Inc., New York, NY, USA.

Index

9 780367 620608